"LUCI"

A novel
By
Rus Morgan

This book is a work of fiction. Places, events, and situations in this story are purely fictional. Any resemblance to actual persons, living or dead, is coincidental.

ISBN: 1-4033-5497-9 (e-book)
ISBN: 1-4033-5498-7 (Paperback)
ISBN: 1-4033-5499-5 (Dust Jacket)

This book is printed on acid free paper.

1stBooks – rev. 4/25/03

ACKNOWLEDGEMENTS:

Writing is a solitary endeavor and the eventual emergence of a finished work is made possible by multiple benefactors. If one tries to thank them all one would start with the dimmest ancestor and the genuflecting would never end. As a consequence, one thanks those directly upstream and hopes those not specifically listed will understand. Under those conditions I would be sorely remiss to let this book fly without paying homage to Sergeant Morgan, Miss Linda, and Michela and the rest of the staff at the VA Hospital in Memphis who work their miracles on me from time; my wife Vonn who absorbs my ire when I am out of sorts; that sterling proofer, my Son Brad who monitors my pros and cons and my old friend Morris Katz who daily teaches me the true meaning of fortitude and perseverance.

PROLOGUE

In a death row cell in The Florida State Prison a surly mass murderer who mowed down fourteen teenagers and the manager of a packed fast food restaurant flipped through the pages of a popular women's magazine. He paused where the fresh faced model was selling the perfume, opened the sealed flap to let the fragrance fill his cell, then tore a hole in her pouting red mouth and jacked into it.

* * * * *

On the west side of San Francisco Bay a triple axe murderer on death row in The California State Prison at San Quentin who has spent his last twelve years and two million of the tax payers money piling appeal on appeal receives in his junk mail a women's magazine. In it is one of those tear open perfume ads with a smidgen of the smell of the latest offering from a prominent perfume maker. He inhales the delightful fragrance deeply, mean while lost in the erotic memories it triggers.

* * * * *

Roughly a hundred miles to the east in California's Folsom State Prison a serial killer who froze his victim's livers then thawed them out and fed them to his unknowing friends on holidays is in a holding room in the middle of a discussion with his lawyer who is writing his eighth nuisance appeal trying to stave off the inevitable. The lawyer has delivered his client's mail in which is the latest copy of a popular women's magazine. The prisoner idly leafs through it while making lewd remarks about each female picture. At one advertisement he stops, kisses the model's face then pops open the fly leaf and inhales deeply of the latest fragrance in the perfume line she is hawking.

* * * * *

Half way across the nation in the Joliet Correctional Center a poor Chicago boy who turned himself into a nationally recognized

Satanist by sacrificing at least ten runaway children reclines on his bunk massaging himself as he flips the pages in the same popular female magazine. One double truck is an advertisement for the latest exotic fragrance. He tears open the flap and inhales deeply with great satisfaction.

* * * * *

In The Mountain View Unit Women's Prison in Gatesville, Texas the madwomen who carved up her lesbian wife and two adopted baby daughters and fed them to the disposer is day dreaming on her bunk as she leafs through the latest issue of the same popular women's magazine. She pauses at one of those perfume advertisements, tears open the glued portion and greedily inhales the luscious fragrance released into the air. She lays the spine of the book against her pudenda and masturbates quietly.

* * * * *

Four Days Later

The madwomen in Gatesville, Texas caught what appeared to be a common cold. At first her nose began to run copiously, at which she complained and the unsympathetic guards gave her a box of tissue. Suddenly she complained of being hot. The female guard took her temperature which was 101 so the doctor was informed. It was easier to get the doctor to her than her to the doctor so the doctor made a house call. But by the time he arrived the mad women was acting like one, screaming in pain and coughing mouthfuls of her own blood. Her lungs pulled it in like a giant sponge and as the doctor worked frantically on her she drowned.

* * * * *

Florida suffered through an abnormally early cold spell. After coming back from one of his daily walks the mass murderer came down with the sniffles. He requested a trip to the infirmary where the doctor examined him, noted significant chest congestion and suggested he stay overnight. The mass murderer took the prescribed medicine but declined saying he wanted to sleep in his own bed. The guards returned him and put him to bed where the Doctor's sedative took effect and he went to sleep immediately. He did not answer morning call and was found dead in his bed. Autopsy disclosed substantial degeneration of the lung tissue accompanied by massive hemorrhaging. He had drowned during the night in his own blood.

* * * * *

The Satanist despised medication and refused help when he caught a cold. He vowed the Black Angel would cure him and he plastered himself against the back wall of his cell in his best imitation of Lucifer standing at the gates of Hell. This lasted perhaps twenty minutes until the intense, searing pain he was feeling in his lungs combined with the bloody fleck on his lips made him scream for human help. Humans were unable to stem the bloody flood in his lungs and he succumbed under the Doctor's puzzled eye. Autopsy disclosed he died of drowning in his own blood.

* * * * *

The axe murderer finished up a new brief for his appeal when he noticed he had the sniffles. It was annoying because whenever he lowered his head over his key board his nose emptied into the keys. Suddenly a searing pain roiled through his lungs. A pain so devastating he couldn't gasp for breath, let alone cry out. His open mouth sucked air and the air rattled through the free flowing blood coursing through his alveoli. He vomited blood all over his word processor and pitched forward into the mess. When they found him at bed check he had been dead over two hours, drowned in his own blood.

* * * * *

The serial killer was kept in non-contact isolation simply because when he got hungry he tended to consider any warm blooded animal good to eat. He had taken his normal exercise period and read his mail which included a couple of popular women's magazines and had retired under the watchful eye of the guard watching him on the TV monitor. He smoked a gift cigar and lazily contemplated how good the guard's liver would be sautéed with purple onions in a sauce made with fresh churned cow butter. The cigar was a good one and had come in the mail from one of his many admirers. Funny how many people admired a man who could do the unthinkable, he thought. He got letters all the time asking him how it tasted, some even wanting his recipes for different parts of the body. Some of those people were crazy though. They wanted to cut up the parts and burn them on alters. All he wanted was a square meal and he particularly liked fresh liver.   He finished the cigar, stepped to the toilet, dropped it in, relieved himself and flushed the John. As he turned he felt movement in his chest. A definite movement like something maybe as large as a grapefruit had shifted position. Suddenly the pain hit him. A searing, wrenching, burning pain unlike any other he had ever experienced. It drove him onto his knees in the corner by the lavatory. He opened his mouth to scream but the pain sucked the sound back into his throat and buried it in the middle of his chest in a liquid, gurgling groan. He opened his mouth to suck air again and blood gushed from his lungs, flowed down his chin, onto his chest and the floor. He pitched forward into the corner. Fifteen minutes later the guard, following procedure, sounded the alarm that signified he had not seen the prisoner for at least fifteen minutes. They found him next to the john hunched over a pool of blood, dead -- drowned in his own blood.

CHAPTER 1

Wardens, Assistant Wardens and jailers et al are a breed set apart. They belong to a closed society with unique problems, long hours, rare vacations and they generally believe that no one but another jailer understands what they do. Hence they tend to be herd animals, corresponding, collaborating and sharing information and experiences. There are a few mavericks that walk the periphery alone and a few who pattern themselves after the Marquis de Sade but most of them are honest folk trying to do an honest job.

Television has brought them into the 21st century. Four Saturday mornings a year "The Warden's Fraternity" meets via teleconference via scrambled satellite signal. As they all are, this one is chaired by "The First Associate" Jed Foley, Senior Warden of New York's Attica Correctional Facility, who is the founding father of the conference. The 'chair' is elected until unseated but no one would seriously think of deposing the popular Foley so he plows on.

It costs money to set up the teleconference and each is charged a fee. To participate one must belong to the 'fraternity'. 'No tickee, no washee" If one doesn't pay the fee there is no connection to the conference and the elevated fee self-limits the attendees to a maximum of twenty five seats.

The signals for the conference pass through a two stage scrambler and without the de-scrambler the communication is gibberish. They exchange ideas, choose officers, define problems, discuss solutions and projects. The conferences get detailed, graphic and sometimes very heated.

These teleconferences are unlike any other conferences any of the members have ever attended. Using modern electronics each attendee enters a small, sound proofed, booth in a corner of his own office, sits down in a comfortable chair and dons special headgear and a pair of light, flexible gloves. As He flips the switches on the console in front of him inside the headgear a large, corporate boardroom lights up and he appears to be seated at a round, elegant, oversized mahogany and leather table. He is apparently joined there at the table by each of the other Wardens as they come on-line. The system is state of the art and the perception is totally lifelike although they are operating in 'virtual reality".

CHAPTER 2

09:00 Saturday Morning EST, 14:00 GMT

"The First Associate" gazed around the table, his dark eyes all but lost between the beetle brows and the pug nose. A Neanderthal jaw lends his head a simian cast. A head which is an obelisk connecting two muscular shoulders and sits on a neck so short as to be practically nonexistent. With grudging respect his inmates and most of his colleagues call him "The Gorilla". A big, rough cut man, with a small paunch fanny-pakking below his barrel chest. His large capable hands once did the manual labor that put him through college and on to law school and first in his class. In early adulthood he married a comely but timid young lady from the old country and in their forty plus years of domesticity he has never strayed from her hearth. He dislikes three things: politicians who forget their constituents; prisoners who try to beat the system and most of the rest of the human race. He is as tenacious as he looks and is well known in Washington circles. As a venerated member of the President's Panel on Crime his phone calls are cut through to the Oval Office as they are made.

He searched the room carefully. "I see that everyone is here, shall we begin?" He appeared to consult a sheaf of papers in front of him then looks up. "As a matter of protocol I'll ask Mr. Secretary to read the minutes of the last meeting."

Paul Ashton, Warden, Folsom State Prison, cleared his throat. He had reluctantly accepted the position of Secretary and only after Foley had made a personal appeal. He was not comfortable with Parliamentary Procedure nor had he yet been able to relax when wearing the electronic equipment. Forty five years old, twice divorced and now single and a veteran twenty-five year correctional officer with the last four as Senior Official at Folsom he was not an imposing physical specimen. In fact, with his receding hairline and pudgy stature, in a crowd you would tend to slot him into the category of 'bean counters'

3

He leaned forward. "Give me a show of hands if you received the faxed copy of the minutes of the last meeting?"

All raised their hands on high. "In the essence of time the Chair assumes if you have received them, you have read them. Also, in the essence of time, you have had ample opportunity prior to this meeting to dispute if you were so inclined. Consequently, I need one hand raised to place the motion on the table to accept as faxed (a hand was raised), Thank you Warden Colbert -- I need all hands raised to accept the motion. (All hands were raised). Thank you Gentlemen. Mr. First Associate the previous minutes have been unanimously accepted and noted into the record. I relinquish the floor to you."

Foley sighed audibly and spoke. "Gentlemen, I've been in touch with most of you during this past week but I'll recap those conversations for those of you who are to date unaffected. We really have only one topic here this morning."

The room hushed and the attendees leaned forward expectantly.

He went on. "During the past five days, five death row inmates in five different prisons have died in their cells -- all from drowning..."

A ripple passed through the room. Several voices at once called out, insistent, "Drowning...come on Warden, ain't no way...not in my cell, under guard..."

He raised his hand and the room went silent. "Yes drowning, drowning in their own blood!"

Most of the room was stunned except for the five wardens in question who looked at the rest of them and nodded verification of the Senior Warden's bombshell.

A chorus went up. "Why haven't we been told about this...suppose it had come out in the papers?...how could anybody reach them on death row?..."

Above the uproar Warden Foley managed to make himself heard. "Because we are family we've managed to keep it quiet up till now but for obvious reasons we can't keep it out of the media much longer. As we speak each one of you is receiving copies of the autopsies, the final logs, the investigative reports and photographs. Those are the facts, and the fact is that they are dead. We know how they died, even when they died. What we don't know is how some body got to them. Now I'm going to flash some of the gory details on the big screen so you can come up to speed on this thing."

The room dimmed and one end turned into a screen on which a series of gruesome slides ticked off. The room was hushed except for an occasional soft expletive, "Jesus...Goddamn...Is that shit catching?...lookat all that fuckin blood..."

The slides ended and the lights came up brightly and sharply. A voice spoke, "Gentlemen, may I have your undivided attention!"

It was a deep voice, a commanding verbal thrust deeper than a fog horn booming off Lands End at 4 a.m. in heavy fog warning ships off the point. A voice so deep that it shivered the bones in your ears, made imaginary icicles skitter across your shoulders and brought you up short gasping for breath. Afterwards they realized the voice was so deep as to defy human reception but somehow every one in the room understood every word he subsequently said.

He appeared in an instant – if the devil can be called a he. Appearing from within a blinding light he towered nearly eight feet tall, massive in arm and chest, with a long, sinewy, serpentine neck as thick as an ordinary man's thigh. The great, corded neck presented a malevolent preying mantis head that turned this way and that as he glowered at the room. Red from head to toe, almost a fluorescent glow about him and if you studied him, Ashton later reflected, you got the impression you could see right through him. Two curved, glistening black horns protruded from the top of his bare skull and his eye sockets, each as big as a coffee cup, enclosed no eyes but each threw out a beam that glowed with the consuming intensity of a small sun. His mouth, set in a perpetual, ghoulish rictus surrounding blood red shark's teeth emitted a deep, rumbling chuckle and as he turned his hideous head to survey the men at the table each man in turn was swathed in the twin rays of evil thrusting from the eye sockets and involuntarily shrank from them. He stood, legs splayed, ham sized hands braced on his hips, three inch black scimitar talons at the end of each of his fingers clicking his impatience. Now around his neck, now about his horns, again around his waist his python sized tail constantly writhed and thrashed – the point of it never resting.

He spoke, or rather began to articulate, the voice octaves below the lowest tuba. A deep, guttural, visceral growl. The table resonated with it, the chairs they sat on vibrated to it, it ate into their chests and arms and they lacked the will and the audacity to move.

As he spoke he moved fluidly, impatiently back and forth between the table and the end wall. All turned to watch him. "Today, I bring you a message of life or death," he growled, "do not interrupt me as I speak. It is your choice whether you live or die. By my presence here and the deaths of the condemned in your prisons I have already proven to you that I can reach into your innermost cloisters." He paused for heavy emphasis and faced the table. "Those deaths were but a sample. I leave it to your scientific community to decipher what killed them and the method of delivery – and the trigger as they are not concurrent. But it will be to no avail. I can infect anyone, anywhere in your world and then trigger the response at my leisure. There is no defense, no cure -- you are all vulnerable – and I have more planned in many high places." He added this last pronouncement off handedly, almost lightly as though as an afterthought he had just decided to throw another piece of wood on a brightly burning fire.

"What do you want from us?"

The gruesome gargoyle lifted his head, casting about for the human who had dared to speak. He finally pinned a wary Foley with those awesome probes. "Ah, Mr. First Associate, it is you who dares break the silence."

Foley winced but gained confidence, "You must want something?"

The Devil cocked his head and mocked. "One of the first laws you humans learn, eh? Every thing is for sale, everyone has his price. Aren't you lucky to inhabit a world based on greed. It makes it so easy to read the bottom line and everyone understands it when you do." His voice changed, turned hard, "What I want is what you puny parasites place the most value on, that which you have been amassing for me for the past 200,000 years. The wealth of your world is mine, Mr. First Associate. I want all the power and energy your world can muster. I want it under my thumb so I can assign it goals and dictate its use..."

"How?..."

The monstrosity flashed that evil shark's grin. "Coin of the realm is the key to the power, Mr. Foley. I will systematically collect your world's wealth and when I have it I will let you destroy yourselves because you can't live without it. As a beginning..." Lucifer splayed his giant hand out to the captive audience and began ticking off his demands, "...I will start with the United States and I will start with one trillion of your dollars...."

"One trill..."

Lucifer cut him off snickering. "Yes, Mr. Foley. One trillion, but that is only a down payment on your world's wealth. In forty-eight hours I want the first installment on the down payment. That amount is one hundred billion dollars. You will amass the credits in two different banks and the money is to be on deposit precisely at 10 p.m. GMT Monday night. You will place the account numbers on your Global Internet at LUCIFER using my password which I will supply to you later. I will have the money transferred to its destination at my leisure."

"I have no authority to ask the president for that kind of money."

"Trust me, Mr. First Associate. By the time you get to your president he will be quite willing to embark on any program that will stop the bloodshed. As will the rest of the nations on your world since I have already impregnated the heads of state of all your nations. Now all I have to do is deliver the trigger at my leisure and I will do that often enough to keep my wishes uppermost on your President's agenda."

Foley sucked in his breath, "Who are you really, and why are you doing this?"

The monstrosity smiled and his evil rumble filled the hall. "I am Satan – the Devil -- the rogue angel – Lucifer, if you will." With a satisfied smirk he continued, "Since I am man and woman you can call me Luci."

He turned his back on the assemblage, stood at the window as though gazing out then swung back to them. His eyes blazing with hate. He raised a massive arm and jabbed his finger at the assemblage. "I have watched your species since you first evolved. You have polluted everything you have touched including the very air you breath. Everything your kind has accomplished has been through trickery, greed, deceit, chicanery and deviousness" He laughed and went on, " In each of your religions you have created me and now I come as your worst nightmare to haunt you. I am the putrid core of your nature. I will take you to task and when I have done with you, if you are still alive you will be back in the woods climbing through the trees looking for stray fruits."

He laughed a long, low grumbling, rumbling belly laugh and began to fade from the scene.

"Wait! When will you come back again?" Foley yelled.

Luci paused. "Gather your nations at 10 p.m. your Greenwich Mean Time two days hence on Monday. I will address them by teleconference as I am addressing you..."

At this moment Warden Joseph Herrin, Florida State Prison, on the opposite side of the table from Luci stood and roared at the top of his voice. "Hey freak, you're nothing but a loud mouthed picture on the wall. By what fucking right do you come in here and tell us what to do..."

Luci stopped disappearing, came momentarily back into full focus, his eyes seeking out this accuser. As the twin probes settled on him, Herrin stopped, uncertain, glanced at Foley then turned his attention back to Luci. He opened his mouth to continue his tirade but stopped in mid-word as Luci raised one gigantic arm, stabbed a sausage sized finger at him as though to merely scold him – but there was a thunderclap – a fracture of the air between the devil and the spellbound warden – a fracture through which a ruby red bolt of pure energy snapped from Luci's extended talon and seemed to pass straight through Herrin's head. Herrin dropped forward onto the table and did not move.

Luci brought his rigid finger to his ghastly lips, blew across the tip of it as though he was clearing smoke then made as if to shove it down into an imaginary holster at his right hip. He chuckled, "Awesome, isn't it?" He surveyed the room and began to fade again behind these parting words, "Read the handwriting on the wall humans. It says you are mortal. Do not mock me if you wish to stay alive a little longer."

* * * * *

Paul Ashton sat transfixed. He was next to Warden Herrin and the bolt that had struck Herrin had sizzled through the air. It had electrified the air around him and made him tingle like a thousand needle pricks at once. He stirred and reached to Herrin who was slumped over the table. It was the first time Ashton had ever tried to touch someone in virtual mode. There had never been any intent by the producers to make it possible to physically intermix and it was a strange sensation to see his hand drop down through Herrin's arm. He could not disturb Herrin's body space and there was no response.

Foley spoke. "How is he, Paul?"

Warden Ashton looked over at Herrin's head. He gasped. "He has a neat little hole burned through his head just above his ear. If his real body is this way he's dead."

The group broke amid a bedlam of voices. It was only to maintain procedure that Foley spoke the words of adjournment as most of the Wardens had already broken contact in a flurry to clear the scene in case Luci returned.

* * * * *

CHAPTER 3
SATURDAY MORNING 11:00 EST, 16:00 GMT, − 52 hours

"What's your next step, Jed?"

"The President's personal chopper is going to pick me up in an hour and drop me at the White House. I've got the password now. I got a phone call a few minutes ago from someone who identified himself as 'an agent of Lucifer' and he gave me a twenty-five digit password for the money."

"Whoo! I wouldn't want the responsibility for that."

Foley was grim. "He told me if anybody else found out what it was me and my family were all dead."

Paul Ashton wiped the perspiration from his forehead. "This thing gets really scary when you realize that we were only looking at a virtual world. Even there this thing can kill. And it kills without remorse. Herrin really didn't do anything that warranted killing him. Do you really think it is the devil?"

Foley shrugged his bulbous shoulders. "Shucks Paul, You religious folk will have to come up with the answer to that one. I'm an atheist so it's a moot question for me."

"Jed, in an atheistic world you can't have a devil because you don't have a god. If there's no God and no devil how can there be such a thing as good and evil."

Foley grunted. He did not like religious discussions, they were an invasion of his privacy. "God and the devil, good and evil are all fabricated in men's minds. My world is clear and concise and I don't have to worry about all that other baggage. Simply put, if I am good and moral it is because it furthers my cause."

"Could you be bad and immoral conversely?"

"Sure but there's little profit in it."

"Jed, I can't believe that. You've always led an exemplary life and your morality and integrity are beyond question. You mean that all came about just because you felt it was the most profitable."

On the other end of the line Foley smiled. "Certainly Paul. That's simple proof that you don't need a flawed society to know where the mile markers are." Foley glanced at the wall clock. "Enough of this, the chopper's on the way, I've got to get ready."

"Wait a minute Jed. What I really called you for was to ask your advice on something. One of my benefactors here at the works is a bona fide genius. An odd old duck with a brilliant deductive mind. He bestowed our new library. I thought I'd give him a call while you were talking with the president and see if he can give us any insight on this problem. He's also a whiz at electronics."

Foley was anxious to get loose. "Sure, why not. If we come out of this we're going to need all the help we can get. I'll call you from DC."

CHAPTER 4
11:30 EST, 16:30 GMT – 51.5 hours

Jason Amador was not a large man. The rich shock of nearly white hair made him seem bigger than he was but also made him look older than he was. In his early sixties he was a slender, fit, six footer. The lightness of his hair intensified the olive complexion and made him appear darker skinned than he was. The hair played off his intense green-gray eyes and gave them a glow that seemed to start back inside his head and pierce you when he looked full at you.

He lounged comfortably like a grateful lizard stretching under a benevolent late morning sun. Below him a serene panorama unveiled a succession of blue green hills looking like shepherds nudging their secluded valleys through the purple haze toward the lower pastures on the plain far below.

He was on the large, main deck of his North Georgia mountain retreat, L'Aerie. A converted grist mill, it stood beside and over a clear, cool mountain stream harboring record breaking rainbow trout.

He had bought the abandoned three-story building and the surrounding twenty acres for pittance and genius and five million dollars had converted it into a secluded, impregnable paradise. Inside the weathered boards of the old mill was a trove of electronic marvels and state-of-the-art computer wizardry that only a multimillionaire could afford and only a genius could put together. At the heart of the complex, humming contentedly in the concrete basement dug deep into the granite hill was a voice-operated, tri-mode computer which would have made NASA envious were they aware of it. Its calculations were so hot and fast that it required the constant flow of the cold mountain stream through the underground scrubbers to keep it cool. Affectionately dubbed 'Cray' in deference to its original, primitive creator, it responded to only two voices and quietly maintained surveillance over security, communications and other life-support systems.

Cray spoke in a soft, modulated, thoroughly pleasant female voice which broke him out of his pleasant reverie. "Mr. A."

"Yes, Cray?"

If anyone were watching Amador would appear to be talking to himself.

"Mr. Paul Ashton is online."

"Thank you, Cray. Patch him through."

There was a tiny click and the warden's harried voice reached out desperately across the miles. "Jason, are you there?"

"Yes, Paul, nice to hear from you."

"You too, can you scramble this line?"

Jason raised his eyebrows. "Yes."

"Please do."

Jason said, "Cray, scramble this line please."

In a moment there was a subtle alteration in the almost inaudible hum omnipresent on telephone lines. "Done, Mr. A."

"Thank you Cray. Paul you can talk."

Ashton went straight to the grist. "Good. Good. Jason, forgive me for letting so much time fall through the cracks since the last time we talked and also for not dwelling on that now..."

"Done."

Ashton cleared his throat. "You remember our teleconferences every three months?"

"Yes."

"Well, this past one leaves me with as bizarre a story to tell you as anybody has ever been asked to believe but believe me I am sober and I am not crazy..."

The story tumbled out scene after scene until Ashton had brought Amador up to date.

"Where are you now?"

"My office."

"Good, are you able to come here?"

"Yes."

"Hold on a minute." Still on line Jason spoke to Cray.

"Cray."

"Yes, Mr. A."

"Tell Christa to listen to the tape of this conversation. Contact Mr. Chaney and deliver this message. 'Peter, leave now for Carmel

Valley and fly to SAC. Exec.[1] Embark Paul Ashton and fly to Le Tourneau Field, Toccoa, GA. Chopper will meet you. Paul will fill you in. Confirm to Cray. 'Cray' Confirm to me when that's done. Paul, what size is that video tape you have?"

"8mm DVT."[2]

"Marvelous. Drop it into your answering machine. When Cray dials you switch it to Fast Forward and we'll suck a copy of it back here in about ten minutes."

Ashton's tone was incredulous. "You can do that?"

"Yes.

"Good Lord, the world is passing me by."

"No, just a piece of it. I'm going downstairs, Paul. When Cray breaks the connection that means we have a verified copy. "

Cray broke in, "Mr. A., Mr. Chaney confirms"

"Thank you Cray. Paul, I want you to go to the Sacramento Executive Airport and wait in the gold room. They'll be expecting you. There is food and drink there. Wait for Peter. You remember Peter Chaney don't you?"

"Yes."

"Peter is my strong, right arm. He will meet you and the two of you will jet out here and the chopper will meet you in Toccoa. It's 9:30 a.m. your time so we'll expect you for supper. Bring your original tape, talk to no one, except I want you to bring Peter up to speed on this while you're in the air, and make sure you have your c-fone so I can call you wherever you are. Ta."

Jason did not wait for Ashton's retort but was already rushing through the bullet proof sliding glass doors leading off the deck.

* * * * *

He burst into the basement CPU to see Christa Bonamay sitting at a keyboard, her graceful, brown fingers lightly caressing the keys. Periodically she glanced up at a 4' x 6' wall screen on which a series of numbers was phasing as though Cray was seeking a prime number. Occasionally she touched the small screen in front of her and

---

1Sacramento Executive Airport

2Digital Video Tape

14

interrupted the flow of information. She was surrounded by electronic works all quietly humming. The room was filled with an almost sound like the omniscient whisper one hears throughout the length of a cruise ship.

"You heard the conversation?"

She did not look up. "Yes."

"How are we doing?"

"Cray is in the middle of the download now. I'm decompressing as it comes in."

"Cray."

"Yes, Mr. A."

"Connect me to the Sacramento Executive Airport Gold Room please."

As Christa continued clicking her keyboard a strange voice came through the room speakers.

"Hello, Gold Room. Cassidy speaking."

Jason answered. "Mr. Cassidy, Codeword 'Condor'"

The response was cool. "Just a moment please." A few moments went by and Jason knew that Cassidy was pulling him up in the computer.

"Condor, what are your double letters?"

"A."

There was decidedly more warmth in the voice this time. "What can I do for you Mr. Amador?"

"I am expecting Paul Ashton, A-S-H-T-O-N, to wait for my jet. Please provide for him."

"Like he was you, Mr. Amador."

"Thank You, Mr. Cassidy." Jason turned to Christa who was sitting with her hands in her lap, waiting for Cray to stop humming. She leveled her startling blue eyes on Amador.

"Somebody out there is in front of SOTA."[3]

"Intriguing isn't it? Somewhere out there, some one has made a quantum leap in holography. Imagine, an image that can shoot lightning bolts, wow!"

* * * * *

---

[3]State -- of -- the -- Art

They played the tape, over and over, enthralled by the spectacle. But once they had got past their awe Jason settled down to some serious diagnosis. The first anomaly was an inability on Cray's part to pick up Luci. Luci was apparent to all visually but Cray did not acknowledge his presence as an entity.

In response to Jason's question Cray's answer was surprising. "The 'image' you are describing is a figment. It does not exist..."

"But I see it, dammit!"

"...Yes, Mr. A." Cray's tone was maddeningly condescending. "...but I do not. Within the parameters you describe there is only a minute change in magnetic resonance except when the bolt was fired. That charge was 60,000 volts at 50 amps..."

"Good lord, no wonder it cauterized the wound-Cray, where did they get that kind of power?"

"...There was a corresponding drop in voltage in all the transmissions."

Jason pursed his lips. "So Luci didn't bring that power with him, he collected some from every man present, magnified it, and shot it. Amazing! Cray, How about the voice?"

"It is a simple, digital reassembly of a human voice."

"Anybody we know?"

"I am searching now."

"Interesting," Jason said to Christa. "Cray tells us our eyes are deceiving us, that there's only a slight electrical disturbance there and the voice is a human voice delivered digitally. Awesome."

"What is amazing," Christa answered, "is that it is done in real time. Jason, can you calculate what kind of speed and power we're dealing with that can put that monstrosity in that room and have him and his voice react to circumstances in REAL time?."

"Yes," Jason mused. "I can comprehend it but I'm not sure even Cray is capable of that Christa, and how did they get this 'thing' in that room. A secret meeting, scrambled code, everybody on virtual reality and no continuous beam of light."

"Holographs need a light source, don't they?"

"Yes, experimental science shows us there is a way to produce a hologram using just one single source of 'white light', but conventional science is that a hologram is created by a single beam of light originating from a single laser. The beam[4] is split by mirrors and reformed by other means into a coherent beam that displays the image. The process utilizes both the amplitude and the phase relationship of the light reflected from the original object."

"But that's only for a static hologram, isn't it, something like you'd find on a credit card?"

Jason smiled. It pleased him when Christa could follow his line of thinking. "Therein lays the immensity of this project. A credit card has two images and when you manipulate the visual angle you see it phase from one to the other and you get the impression that there is motion when there is not – it is one image flowing into another like flipping the pages of an animator's sketch pad. On the other hand, deforming a hologram in a predetermined pattern, such as a head rotating, or a bird flapping its wings, requires a series of phase changes and these simple visual changes require monstrous amounts of computing power – and they take many minutes. To do this in 'real time' and allow Luci to react to an unknown set of circumstances is literally a scientific phenomenon."

Cray spoke. "There is also a subset of digital information on this tape, Mr. "A"."

"Is it just background noise or does it say something."

"It's music, or to be more correct it is a recurrent beat."

"Anything we'd recognize?"

"No match in my files."

"Why don't we hear it?"

"It is below human perception. A hum so low that if human ears perceived it they would not recognize it for what it is."

"Cray, do you consider the sound significant?"

"No correlation. I mention it because it was present simultaneously with the magnetic resonance. It is important only because it is on the tape."

"Thank You – any information on the voice?"

"Yes. Southern European, probably northern Mediterranean, male between 20 and thirty, public school education through perhaps twelfth grade..."

---

4 See "Compton's Enc., 'Holography'."

"Native language?"

"...Italian."

Jason drummed his fingers on the desktop. "Cray, was any of his syntax programmed and if so how much?"

"Elocution, pronunciation and cadence were consistent with memorized material delivered on cue."

Christa snorted, "What are you saying Cray, that this was an actor speaking lines?"

"Yes."

"Probability factor?"

"Eighty-seven percent."

Jason and Christa looked at each other in astonishment. Jason grumped. "He's nothing but an errand boy, a messenger – a very dangerous, unpredictable servant to somebody – but to whom?"

Jason drummed his fingers on the table for a few moments then shook his head. "Cray, locate every reference to Lucifer on the Net. Give Christa a hard copy of all those that are pronouns and/or are used as addresses on the Net." He turned to Christa. "When you get that copy deduce the twenty-five most likely to be our monster and lock Cray onto their traffic. If we're lucky we can catch him in the act, if not maybe we can at least observe."

* * * * *

Jed Foley quietly submitted to a detailed body search which included his mouth, both ears and his brief case before being escorted by the President's Secret Service Contingent into the spacious interior of the Bell 214ST Twin Turbine chopper. It lifted from the Attica Pad like a giant dragon fly and homed in on The White House with the purpose of a flung javelin.

* * * * *

A short time after Jed Foley lifted off the Attica Pad Paul Ashton stepped off his own chopper at The Sacramento Executive Airport and was directed to The Gold Room. Aptly named, it exuded the deep, quiet hush of an English Gentlemen's Club. A polite attendant deferentially checked him off a list, sat him in a two thousand dollar tufted Chesterfield chair, offered him a menu, took his drink order,

18

returned with it and was off to make him a sandwich all within the first three minutes of his entry into the room. He took a deep drought from his single malt scotch and branch and sank back into the supple folds of the top grain leather. In fifteen seconds fatigue overtook him and the waiter found him snoring softly when he returned with the sandwich. He covered it with a Sterling Silver dome at Paul's elbow and slipped off to other duties.

* * * * *

CHAPTER 5
10:30 PST, 13:30 EST, 18:30 GMT   - 49.5 hours

"Here's the download on the death row autopsies, Jason. Cray says they're identical. In essence everybody drowned in their own blood."

Jason looked up. "Same thing I've been getting here Christa, look what Cray has been filtering off satellite." He indicated hard copies scattered on his desk. "It's been happening all over the world and they are all the same. At least twenty-five confirmed cases and they are showing up almost as fast as Cray can translate them into English."

"Do any of them have any theories as to how the agent is delivered?"

"No. But it's obvious, this monster can reach out and kill at will. Christa, get me Dr. Daniel Ebo at the CDC."[5]

"Mr. A..." Cray interjected.

"Yes."

"Mr. Chaney confirmed he is airborne out of Sacramento Exec as of fifteen minutes ago with passenger Paul Ashton. ETA[6] Le Tourneau Field Toccoa five thirty p.m. EST."

"Good. Cray I have two chores for you."

"Yes, Mr. A."

"Chore Number one. Analyze the circumstances in these thirty or so murders we've uncovered so far. Analyze in two groups – the inmates – and the rest and keep your answers in those two categories. Give me your assessment of the following:

What is causing the deaths?
How is it delivered to the victim?
What is its lifespan?
Where did it originate
Can we detect it?

---

5 Atlanta Center For Disease Control.

6 Estimated Time of Arrival.

Is there an antidote?

Chore Number 2:

A pity you have no eyes to see this Luci monster with, Cray. However, I want you to trace that magnetic resonance and/or the 'hum' back to its source.

Where did it originate?

How was it created?

How is it directed.

and most important, is there anyway to fight this thing?"

Christa interrupted him. "Jason, Dr. Ebo is on the phone."

"Go Cray. Daniel, this is Jason Amador."

A quiet, disembodied, cultured, south African voice came into the room. "Jason, my old friend. How nice to hear from you after all these years."

"Daniel, I've kept track of you and you have surpassed my wildest predictions."

"You are too kind. What brings you to me electronically today?"

Jason smiled. "I always loved your lack of circumspection Dr. Daniel. Wait just a moment, I have a voice match on you but I need a visual. Switch your camera to export, please and let me see you."

In a moment the face of a serene, smiling, slightly graying slender African filled the 4' x 6' screen. His double row of brilliant white teeth matched the whiteness of his lab smock. "Good, you haven't aged a day. Now Daniel I'm going to scramble." The screen jumbled momentarily then jerked back in to focus. "I hate to be so theatrical but you'll understand in a moment. Are you aware of the weird death row murders and at least twenty-five more similar throughout the world?"

Ebo's face went grave. "Yes, we are getting reports recently on the five inmates but what are these new ones?"

"Set up down there so Christa can download to you and we'll both be seeing the same thing. Have you had a chance to develop any theories on the five inmates.?"

Dr. Ebo was about to speak when his printer started clattering and he went to it. As he moved through his lab the camera followed him showing a room filled with rows of shelves holding specimen jars, glistening stainless steel tables and buzzing electronic equipment. He was alone. To Jason it appeared that he was actually in

Ebo's lab. Ebo began scanning the info as the high speed printer churned it out. He turned to the camera.

"These simply reinforce the information on the first five. I'm waiting for more detailed tests but preliminary information indicates the agent is a virulent, short-lived bacteria. It hemorrhages the lungs by simply cannibalizing certain interstitial tissues which leaves the blood free to go where it will. We have no idea how the bacteria was introduced so we do not know whether a few of the bugs have a very big appetite, or whether millions of them did this...

"You said 'short-lived' Doctor. What order of time are we facing here?"

"Apparently two days gestation then a minimum of twenty minutes. In the case of the Folsom inmate he was only out from under surveillance for twenty minutes. In the previous twenty minutes he had been discoursing with his Solicitor and reading his mail and was apparently normal..."

"Any word on how the Attorney is doing?"

Ebo shuffled through some papers. "Matter of fact we do. It says here that he went back to his office and was notified there some two hours later. We have had no word of his being in distress..."

"Do you have any thoughts on how to combat this scourge, Daniel?"

Ebo shrugged his shoulders. "In order to have any hope of reversing the procedure one would have to know the exact time of exposure and one would have to be able to react instantly with the proper antidote. But I must admit that if a patient was brought into this lab right now I would be at a loss as to how to treat him." Ebo turned full to the camera. "Jason, what is your interest in this?"

"Daniel, what you see there is just the tip of the berg. In the next few minutes I'm going to be getting some deductions from my computer. We call her Cray and she speaks and Christa, my assistant, is going to patch you in on the circuit so you can interact. You'll hear some answers that corroborate yours and hopefully, some new ones, but this will serve to bring us all up to the same speed in less than five minutes. We can talk further after we've listened to Cray..."

* * * * *

"Mr. A"

"Yes, Cray."

"I have answers to your questions."

"Good, Daniel, can you hear us?"

"Mercy yes, what an experience, is that really a computer speaking?"

"Yes. Cray say hello to Dr. Daniel."

Cray's svelte voice carried to Ebo. "Hello Dr. Daniel."

"Hello Cray. Heavens Jason, it sounds like I'm talking to a young woman," Ebo said self-consciously, "do I talk to it or to you?"

"If you want to ask her a question simply preface it with her name and she will respond. Cray, you may answer Dr. Ebo's questions during this conversation."

"Acknowledged Mr. A."

"Good. Cray now give us the answers to my questions in the order I gave them please."

"Death in all the cases was caused by interstitial damage resulting in terminal hemorrhage..."

"...nothing new there, Daniel. Cray, how about delivery?..."

"...Insufficient input on the twenty-five. On the inmates the only common denominator is inhalation..."

Dr. Ebo sucked in his breath. "That is logical. That explains why the lungs..."

"...so they're getting it through the air? Cray..." queried Jason. "...Yes." Cray answered. "If you wish only one answer to 'how delivered' then it has to by inhalation."

"What's your probability on that one?"

"Ninety four percent."

"What's the closest percent below that?"

"There is a clump on the curve ranging between thirty-eight percent and forty-six percent."

"Not much room for argument there. Daniel, you got any thoughts on how the bacteria is being presented to the victims?"

"At this stage Jason, No."

"Me either. O.K., Cray, deductions on how it was delivered to the victim.?"

"Insufficient data on the twenty-five at-large murders. However, the inmates were all incarcerated under intense guard and scrutiny.

Therefore delivery had to be affected by or within an established routine..."

Jason winced. "Daniel, have you seen any obvious pattern in your reports on the inmates?"

"Not really. Other than the obvious fact that each was locked up each of them was doing something different. Some had visitors, some had not."

"I didn't see anything there either, Cray, how about the life span?"

"Insufficient data on the twenty-five. No data on the pre-terminal phase with the inmates. Once the terminal phase is launched there is a range of twenty two minutes sixteen seconds to two hours, ten minutes twenty-six seconds..."

Ebo broke in and Cray politely paused until he finished. "This virulence is unheard of Jason, even Ebola takes twenty-four hours to start its work."

"...However," Cray continued. "the circumstances indicate a pre-planned life span or the bacteria cannot live outside the original host."

"Cray, why do you say that?"

"There is no record of secondary infection nor subsequent attack on anyone around the victims after the original host is dead. Supporting that is the fact that with the other twenty-five if the bacteria did not die with the host there would now be a world wide epidemic."

"The point is well made Daniel, we'd all be dying already if the bacteria lived after the host died."

"Precisely," Ebo replied.

"Cray," Jason asked. "where did it originate?"

"It is a man made mutation."

"What? Mutated from what Cray?"

"Escherichia Coli."

"Damn! You hear that, Daniel?" asked Jason and without waiting for an answer from Ebo he said to Cray. "Cray, Why E Coli?"

"Of the known pathogens in my files their profiles are not the same but closest to being similar. Escherichia Coli normally lives benignly in the human gut but causes serious infection and sometimes death in other parts of the body. It is much slower than this bacterium. Both are anaerobic and maturate under warm, moist conditions."

24

"Any thoughts Daniel?"

"Marvelous deduction given the paucity of our information."

"Cray, why do you say it's a mutation?"

Cray's voice was calm, methodical. "Assessing the past progression of Escherichia Coli, it would need one hundred forty-two years more, unchecked by man, to develop to this level of speed and virulence and become sophisticated enough to attack the lungs. In order to reach this level at this time it has had to mutate. It is not logical that that mutation would have happened without the aid of man."

"Daniel, is gene splicing possible on the bacterial level?"

Ebo shrugged. "Certainly it would be possible but, to my knowledge, no one in the legitimate world has done it."

"Well, back to the drawing board. Cray can we detect this bacteria?"

"There is insufficient data to answer."

"Well, there wouldn't be time to do anything but look at the symptoms anyway and they would be self evident. All right Cray, is there an antidote?"

"Not in a practical sense. Massive injections of antibiotics are indicated but would be ineffective within the short time frame and doses of that magnitude would probably lead to anaphylactic shock and death."

"Well Daniel, no magic wand there. Cray, any deductions on origin?"

"No direct link but one footnote..." Cray paused and Jason snorted impatiently. "Well Cray, What?..."

Cray continued unruffled. "The most logical point of origin would be one of the third world countries – in particular Iraq."

"Well, the UN never did consummate adequate inspections of the chemwar stockpiles those people had at the end of Desert Storm so who knows. We should have turned that war machine into a pile of glass when we had the chance. What do you think, Daniel?"

Dr. Ebo smiled into the camera. "At the end of WWII there was a surplus of scientists of all disciplines emitting from the vanquished countries. Many of them ended up in the United States but many of them went other places wherever they could make the good money. It is entirely plausible that some very good research Biologists went into the Third World."

"Thank you Daniel. We've got to keep moving on this so I'll sign off. Its been a pleasure talking to you again. Keep me posted on anything new and I'll do you the same."

Ebo smiled up at the camera. "As always Jason, my friend, I am at your service."

* * * * *

CHAPTER 6
15:30 EST, 20:30 GMT   - 47.5 hours

Foley's two and a half hour chopper ride was uneventful. Lost in his solitude, he sleepily watched the ground unfold below. The four Secret Service men chose to respectfully ignore him except for the offer of refreshments half way through the flight. He declined, preferring to spend the time marshaling his thoughts for the grueling conference coming up in the White House. He began to count the minutes only when they entered Washington D.C. airspace and were accepted on the approach to the White house.

When the chopper finally dropped lightly onto the pad at the White House he was again surrounded by tall men in gray suits. He was searched for the second time, this time with a humming metal detector, then escorted to the Oval Office.

* * * * *

Even though he knew beforehand he was picking up his ex-incarcerator a crush of unwelcome memories pumped through Peter Chaney's mind when he walked into The Gold Room and looked down at the sleeping warden. Ashton had never touched his sandwich and the rest of his drink and was still snoring peacefully, slumped over sideways in the deep bosom of the Chesterfield.

For the ten long years that Peter languished in Folsom Prison Ashton, first as the Assistant Warden then The Warden had eventually commanded his grudging respect. Luckily for Ashton now, he had been an honest, fair administrator. Notwithstanding his respect for the balding sleeper Peter took a vandal's delight in waking him up. Ashton gazed up at the tall, dark ex-con through his sleepy fog, then silently followed him out to the ship where he promptly buckled his seat belt and went to sleep again. So much for that, thought Peter and pointed the ship toward Georgia.

Invisible from the ground, the tiny, twin jet Rockwell Sabreliner pierced the sky at the head of a long, thin silver contrail. In deference to the folk on the ground Peter cruised slightly below the sound barrier and let the swift little ship needle its way through the atmosphere. They were just into the second half of the trip, dropping over the apex from 36,000 feet in a long parabola that would bring them to ground over North East Georgia when he saw movement in the small, convex mirror suspended above the instrument panel. He turned to see Ashton making his way forward. He indicated the passenger seat. Ashton dropped into it, strapped on a seat belt and donned a headset.

The Warden yawned and grinned at Peter. This prisoner had never seemed like a prisoner. Something inside had set him apart. He had never looked like a con nor acted like one. Ashton had always been impressed with Peter's lack of 'prison presence'. Chaney had always reacted almost as if he was stopping off at the prison to say hello to someone else. Ashton spoke into the mouthpiece. "Well Peter, how you been since leaving my fair city?"

Peter enjoyed the irony. "Frankly, Warden, I've never been better. Amador is a good man to work for."

"Yes, Jason is something special. How much time do we have left?" Peter indicated a dial. Ashton looked at it then gazed out the window at the filmy clouds grasping at the tiny wings. Occasionally he could see the shadow of the little jet flitting ahead of them as though hopping from clump to clump of the lower clouds. The earth was a faint blue below. They were alone in a tiny, pointed bubble forcing their way through the freezing ozone. Ashton read the outside temperature and shuddered – minus 34 degrees. An ominous mile marker along an already ominous day he thought, and he began to recount the days events to Peter Chaney.

* * * * *

"Mr. President, Jed Foley."

The president was a tall man, almost as tall as Foley but much trimmer. A shock of brown hair, edged at the temples with enhanced gray, and a generous hooked nose lent him the look of a swooping hawk until he opened his mouth. They called him The Gentle Hawk with that voice, the vibrant voice of a man used to holding his audience in his hand, a man used to taking them to the brink, or to the heights or to wherever from there he would.

He advanced around his desk and shook Foley's hand and indicated a settee.

A door to the side of the Oval Office opened and another shorter, older man entered. The President greeted him. "Jed, You've met the Secretary of State, Mr. Ohrbach." The two shook hands and the President continued.

"Let's go down the hall to the conference room. CIA, FBI, ATF, DEA and Secret Service will be there. From what you say every one is going to have to cooperate on this one."

They entered the conference room. Jed guessed it at 20 by 30 feet and it was decorated as one would expect a room to receive ambassadors in would be. It was occupied by the four senior officials already seated at the oval, mahogany table and each was flanked by aides who sat behind them around the wall. Perhaps eighteen people in all. As the President entered all rose. He bid them be seated.

"Gentlemen, all of your staff have met and most of you have met Warden Foley through common efforts on the Crime Panel. For those of you not familiar with The Warden let me say..." and he continued for perhaps two minutes, outlining Foley's background, accomplishments, dedication and integrity. "...The point of this," He continued, looking squarely at each of the department heads as he swept the room, "is to establish beyond a shadow of any doubt that what you are about to hear is being delivered to you by an intelligent, experienced, seasoned, unflappable officer of the law." He paused and patted his forehead with his handkerchief. "By God Gentlemen, looking back on what he told me on the phone earlier today, and some of the faxes I have been receiving from around the world since then, I wonder if we aren't both crazy, but the least I can tell you is that we have an international madman running around killing – Oh hell, Jed bring 'em up to speed." He sat down -- a thoroughly distraught Chief Executive.

Jed Foley rose to his feet. Simultaneously a red light began blinking on the phone in front of the President. He held his hand up to stop Jed and picked it up. First he listened with disbelief, than awe, and than horror quickly spread across his face and he slammed the phone down. Everyone flinched at the sound then sat there in respectful silence, waiting for The President to speak.

If his short speech on Foley's background hadn't served to establish the serious nature of the meeting his demeanor now did. The Command Staff had never seen the youthful Chief uncertain, but here he was --faltering, his hands shaking, wiping imaginary spittle from his lips. Suddenly he had aged twenty years and sat there a shaky, uncertain old man. Foley still stood, politely waiting for the President's permission to begin.

The President spoke, looking at no one, just simply spoke to the room, his voice a raven's croak. "Miriam Conley was an attractive, bright, young lawyer doing a very fine job for me as my Press Secretary. Most of you knew her, and had face-to-face dealings with her and we all knew her as an honest, conscientious, caring young lady. She – She was just found in the Ladies Room lying face down in a pool of her own blood. Warden," He waved his hand at Jed to proceed, "Warden Foley will show you how she died."

* * * * *

CHAPTER 7
17:30 EST, 22:30 GMT, - 45.5 HOURS

Christa brought the Agusta 109C, twin turbine, 8 passenger chopper to a hover and slid it gently to ground at the Executive Airlink at Le Tourneau Field, Toccoa, GA. She had talked to a surprised Peter Chaney on the radio less than fifteen minutes prior and was expecting the tiny jet to drop out of the sky momentarily. Peter had not been aware that Amador had spent a good part of the previous summer teaching her to fly the chopper and that she was now a licensed pilot. Peter had quietly congratulated her on the air and had given her his ETA. She reflected on this with a warm sense of satisfaction as she watched the ground crew fill the Amador-designed special pod tanks with the extra fuel that extended the range of the 109 to slightly over a thousand miles. Jason had requested the extra fuel on board, 'just in case'.

* * * * *

Peter vectored in on the landing strip and deftly dropped the little jet to the surface. As they taxied to the far end familiar buildings came into view and he could see the chopper off to one side, apparently being refueled. Christa was visible in the late afternoon sun, next to the service man, laughing and waving to them. She wore a form hugging purple jump suit accentuating her long, slender legs and high bosom. She held her helmet in her hand which let her long ebony hair flow freely about her dusky face. A natural born sex machine, Peter mused. I wonder what she's doing for it. She certainly isn't getting any up there on the mountain. Not from Jason – not from anybody I'd guess.

Ashton interrupted his reverie. "Beautiful girl that. Know her?"

"Yes – and no."

Ashton was unbuckling his seat belt. He glanced at Peter and smiled. "You mean you do know her, and you aren't getting any."

Peter winced. "Is it that obvious, Warden."

31

* * * * *

"Cray, put your thinking cap on."

"I don't understand, Mr. A"

"You will," Jason responded picking up a peanut butter sandwich and leaning back and propping his feet up on the counter. He was in the CPU, surrounded by his triple mode, 15 Gigabyte computer creation he affectionately called "Cray". Cray hummed quietly, respectfully, doing a dozen tasks already allotted and waiting for Jason's request.

"Cray, earlier I asked you to research four questions in regards to Luci – do you remember?"

"I cannot forget anything, Mr. A."

"Good, you know the subject now give me some answers. Cray where did Luci originate?"

"The magnetic resonance originated from two separated areas. Calculating with the Global Positioning System one half the signal originated on Lissacratus Street in Athens, Greece and the other came from Phillips' waterfront restaurant in Baltimore, USA."

"Can you tell me which floor at Phillips'?" Jason asked idly."

"Probably the second as the first floor is occupied with kitchen, restrooms and customers."

"Smart ass."

"Is that input, Mr. A?"

Jason sighed, owning a computer that was capable of some reasoning was some times disconcerting. "No Cray, ignore that. Cray, how was Luci created in that virtual reality existence?"

"Unknown. No known technology applies. Current laws of physics do not allow this phenomenon."

"Yeah, well, obviously there is something here that you and I have overlooked or we're looking at some totally new math."

"Illogical"

"Granted, but we have an event – something caused it. Cray, can you tell me how the signal got there?"

"Negative but to produce the image you specified with the real time action you described plus the audio, requires a two hundred fifty six bit bandwidth."

"Damn!" Jason said through the last bite of his peanut butter sandwich. "Nothing operates on that wide a path -- you are unique and you only operate on 128. Cray, are you saying it comes in on radio waves?"

"Probably," said Cray. "Hence the paradox."

"You learning to use some new words?"

"Miss Christa continually upgrades me..." Cray answered. "As of last count my dictionary contains a merged/purged list from Grolier's, Compton's, The American Heritage and Random House Dictionaries; Roget's Thesaurus and the Encyclopedia Britannica plus full medical, engineering and scientific dictionaries. Some 822,576 English language words plus every day conversational capability in 2000 other..." Jason could not detect any change in Cray's tone but could swear he detected a self-satisfied smirk in her words. He cut her off. "O.K. Cray, enough, enough!"

"Yes, Mr. A.

"Cray, back to Luci. Is that signal self sustaining or does it piggy back on an existing signal? -- Wait! Let me clarify that – can that signal come in by itself or does it require an existing signal to ride on?"

"In the data available a video/audio signal was present prior to the entry of the magnetic resonance."

"Is there any data to indicate if an existing signal is primary?"

"Negative."

"Cray, here's the biggie. How can we fight this monster?"

"Standard holography requires a strong, steady source signal. The obvious solution is to interrupt that signal."

"But if we don't know what it is, how do we disrupt it?"

* * * * *

They toted their bags across the tarmac to the chopper. Christa met them and extended her hand to Peter. Her eyes met his squarely. "So glad to see you again Peter, it seems like such a long time."

Peter mumbled something and felt his heart jump at the electric touch of her fingers. They were long, tapered fingers – a piano player's fingers and they were cool against his palm. He held them for a long moment until Paul Ashton chuckled and broke the spell.

"I'm Paul Ashton."

Christa turned her attention to him. She was as tall as he and he was struck immediately by the dramatic, vibrant blue of her eyes. He found it impossible to take his eyes off hers as they talked. She lowered hers occasionally to break the spell but savoring his admiration raised them to meet his again.

"Hello Mr. Ashton. I'm glad you made it safely. Mr. Amador will be very pleased that you've arrived."

Paul indicated Peter. "Please call me Paul. We had a very comfortable trip out, I slept a good part of the way and Peter is a very smooth pilot. How long is the trip from here?"

"About twenty minutes over the hill." The service man handed her a clip board and while she was signing he turned to Peter. "We've been expecting you Mr. Chaney. Give me the keys and I'll take care of the ship for you."

"Treat her gently," Peter said.

"We always have an inside spot for Mr. Amador, Mr. Chaney. We'll check it over, fuel it and it'll be ready when he wants it."

"Thanks."

Christa stood by the Agusta's door. "If you gentlemen will board we'll off into the wild blue yonder. Peter, will you co-pilot please."

Peter grinned and stepped into the ship and over to the left seat. She got The Warden settled in the passenger compartment then joined Peter in the cockpit. They battened down and Christa went through the start-up procedure. Engine temperature was still acceptable so she lifted off immediately and wafted the ship up above the trees and headed west up toward the crest of the ridge.

"You did that ve-e-e-ry smoothly." Peter said as the woods flowed below them.

"Thanks, I get as much practice as I can and I had an excellent teacher."

"Is there anything he can't do?" Peter said, for want of something better to keep the conversation going.

She smiled. "Not really, but then he thinks the same of you."

"And you. "

"Yes, you and I – and Cray. We're his family. We're all he cares about."

They topped the ridge and flowed down through a small valley. "In all the years I've known him, Christa, I've never heard him mention his family, have you?"

"No, it's almost as if he dropped in from another planet. He never discusses anything personal like that. "

"How about you, Christa? Do you ever get personal? What do you do for amusement up there on that mountain on your off hours?"

Christa glanced at Peter. He was feigning interest in the view out the other window but was actually hanging on her answer.

"I cook and I paint and of course, I spend a lot of time on Cray."

"I'd like to see some of your paintings."

"That's no problem. Jason won't let me sell any of them so the house is loaded."

"Are you good?"

Christa was enjoying his minor confusion. In earlier encounters Peter had made many assumptions about her and most were wrong. He was now stumbling over new ground with her and totally unsure how to proceed without first shooting himself in the foot.

"Michelangelo needn't fret, but I am pretty good"

L'Aerie was a pinpoint on the mountain in the distance. Peter turned to look at Christa. "What do you paint mostly?"

She thought for a moment. "Strength and grace are my favorite themes. I've painted Jason several times. He's a perfect example."

Peter caught the undertone. "You love him, don't you?"

She bobbed her head. "From the first day I met him I'd have given him anything he wanted. But he's not about physical love, nor lust, he just doesn't seem to need that. None of those things compute in his world." She sighed. "I'm like his favorite daughter and that's all I'll ever be to him."

"Well, then there is still a chance for old Mr. Chaney."

She pulled the chopper around from the north to bring it into line with the approach to the L'Aerie pad. "Well, Old Mr. Chaney, only time will tell. What changed your opinion of me? You used to think once a whore, always a whore."

Peter winced and picked his words carefully. "That was true – once. But over the last couple of years I've had a chance to watch you, and work with you – and let's not forget that you saved my life last year when I was hanging under the Condor."[7]

"Part of the day's work." She laughed. "How's your shoulder doing?"

"O.K. except when it gets cold then it gets stiff."

Peter let the conversation lag as Christa concentrated on dropping the bird onto the pad.

Ashton had studiously watched the two pilots during this interchange. He was not connected to the radio so was not party to the conversation. But judging from the expressions on both faces Christa was still unattached. His spirits soared. She had sparked feelings from his crotch to his head that he thought had been banked long ago. The lady was gorgeous, long and dusky with a beautiful feline sensuality that brought to mind the great cats on the Savannah.

As they had cleared the second ridge it was obvious where they were going. Only one mountaintop retreat was visible. As they neared it the rustic, old grist mill stood out like a huge, brown post dropped among the sparse pines along the hillside. Before they made their swing to the north he could see that a mountain stream ran alongside the house on the south side and there was a huge water wheel. When Christa completed the swing he lost sight of the wheel but could see the hill and house were capped with a large concrete pad walled on the eastern end by the mountain and on the west by the top story of the mill. Scattered about the hillside were indigenous pines, mountain laurel and other native cover. As they hovered just prior to landing he saw three doors leading into the concrete wall on the eastern side of the hill. Christa dropped the bird gently onto the pad and cut the rotors.

* * * * *

---

7 Read "The Voodoo Vortex" by the this author.

CHAPTER 8
SATURDAY 20:30 EST, 01:30 GMT - 42.5 hours

"Paul, the President has the FBI, CIA, ATF, the Secret Service and the whole State Department working on this. They have contacted Russia, England, Japan, China, France, Germany, Brazil, Mexico, Saudi Arabia, The United Arab Emirates and the Israelis..."

Paul was dubious, "Sounds like the other side is being left out..."

"No, its just that we always have a credibility problem with 'those other people' and they hope that this list will mobilize the whole effort faster. Through these nations they intend to reach all the satellite nations and set up a scrambled, round table, video conference/discussion at 9 a.m. EST tomorrow morning, Sunday."

"Do you think it can be set up that quickly Jed. That's less than twelve hours."

"What choice do we have," Foley grumbled. "This maniac has given us a deadline we can't ignore. And he's still killing people. Reports are coming in from all over the world with the same symptoms, probably two hundred bodies by now, one of which was right here in the White House..."

"Jesus!"

"Yeah, I know. President's press secretary, died in the ladies room. Something else, this just came through."

"What's that?"

"Over fifty percent of the death row inmates are now dead and more are coming in as we speak. No body here has the slightest idea how it's being done."

"I was about to tell you about mine. I've lost five more since I came back here. It's a slaughter."

"Well those bastards won't really be missed but it just shows us Luci's power."

"Have you gotten the faxes we've sent you with the information we've gleaned so far?"

Foley's voice was full of awe. "Yes, Your man there is really something. You say his name is Jason Amador?"

"Yes, he endowed our new library at Folsom."

"I hope he keeps up the good work, we can use all the help we can get. In case your C-fone goes out give me a local phone number there where the President can reach you."

At Paul's request Jason had been monitoring the call and now flashed a phone number and fax number up on the screen complete with note that it would get straight into L'Aerie without interference.

"That's great," Foley said. "None of us knows how this thing is going to end up so we have to be able to stay in touch. I've got to go, the President just came back into the room, I'm faxing you some material. Talk to you later.

* * * * *

Later, the four of them were setting around a work table in the CPU. Christa had rounded up sandwich makings and beverages and they were hungrily attacking the food while Jason handed out assignments.

"Paul, I'd like you to stay on top of the doings in DC. Get all the info you can on this conference. Where, exactly when, any electronic information available. Also, I want you to group fax [8] every Warden you know personally. Ask them for immediate reply on any unusual behavior by the dead inmates, any packages delivered, any visits, any phone calls, what they ate, what they wore, what they did in their final hours. Get an inventory of their personal effects and the items that they had with them at the very last."

"The fax is bringing some of that in now from Foley. I'll have the fax ready for the Wardens in fifteen minutes."

"Good. Peter you're in charge of logistics. Make sure the chopper's ready for maximum flight. Christa will give you a list of bugs that Cray has uncovered in our main security system. Correct them if you can and generally assist Christa if she needs you." Peter nodded, took a long pull at a beer and winked at Christa.

"Christa, you and I'll wrestle with Cray. Here are the first items I want you to uncover. Hack Foley's system up at Attica, copy the original tape for Cray to analyze. Put her on it like you'd give a dog a

---

8 Sending the same fax to a pre -- determined number of people. (A group)

bone to worry. I want every smidgen of information on that tape. Hopefully we may get something more than Paul's tape gave us. Also," He turned to Paul, "I want you to listen to this. I am searching backgrounds and habits of all the principles in this show including but not limited to Paul Ashton, Jed Foley, Joseph Herrin and the rest of the wardens present at the original meeting." Paul screwed up his face and Jason laughed. "Nothing personal, we're just looking for patterns. I can get the Warden's names off the tape but I'll need addresses for them. Let's go to work."

* * * *

CHAPTER 9
Saturday Night Midnight EST, 05:00 Sunday GMT, - 38.5 hours

Paul Ashton was collecting faxes from one of the clattering machines at one end of the room. Peter was outside modifying the security perimeter while Jason and Christa worked in proximity in the middle of the CPU.

Christa jacked her chair around and faced Jason who was focused on another monitor. In the wan light he looked his sixty plus years. He raised his eyebrows when she cleared her throat. She referred to a sheaf of papers in her hand. "We already knew that it was a split signal but Cray now tells me that the original Luci signal came simultaneously from two separate satellites. Lissacratus brought in the voice and evidently the image we saw and the restaurant supplied that low-pitched hum."

Jason took his glasses off and rubbed the bridge of his nose. "Cray still doesn't admit to an image?"

"No. Again according to Cray, neither signal was capable of producing a hologram as we know it."

"There's something I'm missing here. Something keeps flashing at me but I can't sift it out yet."

"In regards to what?"

"The holography. There's a blip in my subconscious that keeps prodding the surface like a festering splinter trying to break out through the skin. Some small thing I've read or heard that might help us zero in on this thing. It's active enough to be an aggravation but just not strong enough to come clear into my consciousness. I can't imagine not being able to recall it."

"Do tell," said Christa dryly.

Jason ignored her mock solicitation. "Cray and I have been dissecting Luci's voice pattern. Luckily Cray can 'see' that. It's interesting to note that there is definite correlation between the fluctuations in the subset and Luci's voice."

"Do you think one creates the other?"

40

"No but I do think both are necessary. Initially it looks like the voice and image may piggy-back on the subset."

"One can't work without the other?"

"Exactly. Why else go to the expense of getting two disparate signals into that room." Jason took a sip of orange juice and put the glass down. "Any word yet on the backgrounds?"

"Yes. Warden Herrin was in the midst of a nasty divorce. His wife ran off with the son of one of his best friends. Ashton was accused of cheating on his bar exams years ago but the charge was brought by a classmate who flunked and was dropped for lack of evidence. He's been divorced for ten years and no significant relationships since. Foley's father brought his wife and family to America from Greece when the Warden was three and changed the family name from 'Foulakis' to Foley to make it sound more American. Foley married an American girl of European ancestry. Her maiden name was Anna Marie Martin."

Jason's feet were propped on the desk, his leonine head slumped forward, eyes half closed, his chin almost on his chest. He stirred wearily, "Nothing there, any financial problems – and how about the rest?"

"Three of the other twenty-two have declared bankruptcy in the past. But their fortunes seem to be on solid ground now – and everyone else appears to be stable. Their lives display all the ramifications of the American scene but no significant correlations, coincidences or connections."

"Mr. A." Cray's soft voice interjected.

"Yes, Cray."

"I have the deductions from the accumulated data in the one hundred twenty eight faxes Mr. Ashton received."

"Thank you, Cray. Paul, come over here please, you need to hear this."

Paul pulled up a chair, grinned at Christa and offered Jason the pile of papers he carried. Jason shook his head. "It's all been given to Cray hasn't it."

Paul nodded and Jason indicated the desk top. "Cray," he said. "What conclusions have you come to about the faxes? Verbalize your answers but also put them on the screen."

Ashton could almost imagine Cray stopping long enough to clear her throat before speaking. Her tones were measured, unhurried, her words perfectly enunciated. "The universe here is of one hundred twenty eight individuals in the following diversity: ninety-four percent male, six percent female. Of the males fifty-four percent black, thirty-two percent Caucasian, twelve percent Hispanic, five percent Asian, six percent other..."

Jason cut in impatiently. "Cray! I want significant correlations not stats."

Cray was unruffled. "The significant anomalies in random order are as follows: All victims who died in the daytime were fully dressed; All victims who died in bed were dressed appropriately; All victims drowned in their own fluids; There were no multiple deaths; Ninety five percent of the victims were between twenty-one and fifty-five; The number of females to males is negatively disproportionate; Final effects included five different magazines, toiletries, makeup, combs, miscellaneous erotica and other miscellaneous personal items such as pictures, ID cards, lottery tickets, etc."

Cray paused and Jason looked at the other two. "See anything significant?" They shook their heads and he agreed. "Cray, what is the world wide death count as of now?"

"Five hundred forty two confirmed, another two hundred sixty one suspected."

"Cray, is there any significant information in those numbers throughout the world?"

"Affirmative. The first two hundred eighty-five were in the United States. In the next two hundred half were in the United States and the rest were scattered through the Caucasian world. Of the remainder they are now happening world wide with the least number being registered on the African Continent."

"It started with English speaking peoples first..." said Paul.

"...Then to other Caucasians..." added Christa.

"...Then to people who can't speak English or read." finished Jason.

"Cray, " Jason snapped. "Get me the publisher's names and see if there is any correlation in the death row population with those magazines. Christa, when Cray gets that info tap into the publisher's files and let Cray merge and purge the wholesale mailing lists."

Cray's answer was instantaneous. "The list is ready Mr. A. Of the one hundred twenty eight deaths on death row no one read all five magazines, ten percent read the same four magazines, twenty-two percent read the same three magazines, fifty-one percent read the same two magazines and fifty-six percent read one magazine."

"No help there either, Mr. A." said Peter who had just come into the room and pulled up a chair between Christa and Ashton.

Jason shook his head in resignation. Then he had another thought. "Cray, is there one of those magazines that everyone read?"

"Negative."

"Damn!

In desperation he asked. "Cray, are there any two of those magazines that were read by one hundred percent of the prison victims."

"Affirmative."

"Bingo! -- Well Cray which magazines?"

"'Verite' twenty-six percent, 'Young & Lovely' seventy-four percent."

"Do you think that's significant, Jason?" Paul asked.

"At this moment it's just a correlation which is something we didn't have two minutes ago."

* * * * *

"Jason, I've got the information on the magazines you wanted." Christa spoke over her shoulder as she watched the huge screen above her scroll rows of binary digits.

"How soon will Cray have it downloaded?"

She shrugged. "Soon but there's a secondary step. The Publishers didn't do the mailings themselves. The layouts went from the publisher's computer to the printer's computer and when the printing was done the completed mags were shipped to a mailer who actually put them into the mail."

Cray broke in. "Mr. A. There were one hundred twenty-five thousand "Verites" mailed and five hundred eighty-two thousand "Young and Lovelys". I have compared the two mailing lists and there is a five point four percent correlation."

"A lot of good that does us," snapped Jason. He turned to Christa. "Anything else?"

She smiled. "An interesting point is that the first death row fatality was reported just five days after the first mags were mailed out. Also,..." she paused and consulted a sheet of paper. "...there were four thousand references to Lucifer on the Net but only one hundred twenty-five were proper names. I've got Cray locked onto the twenty-five most likely but suddenly the Net is logging twenty-five new Lucifer's an hour..."

Jason laughed and stood up. "Trying to get in on the loot. On that I think it's time we all grab some zees. Cray will keep accumulating information while we sleep. Peter, will you show Paul his quarters. Let's all be in the kitchen at Oh six hundred for a quick breakfast then we'll have time to tap into the electronics on that nine a.m. meeting up in DC.

* * * *

CHAPTER 10
SUNDAY 8:30 EST, 13:30 GMT   -30 HOURS

"Mr. President, Sir, we need to seat you and Mr. Foley to set the lighting and get a sound check."

The technician indicated the two straight backed but comfortable chairs on the dais. They faced a cluster of microphones and beyond them the cameras. Behind the cameras on the far wall of the recording studio was a bank of 31" TV screens on which other participants in the teleconference would be displayed. It reminded Warden Foley of the "Hollywood Squares" set. At last count eighty nations from the world's current count of one hundred ninety two were logged on and would interact.

Foley watched with detached interest as the technicians miked the President, took lumen readings and adjusted the light. His personal makeup lady dabbed about him, straightening hair and masking his shiny nose. As they sat quietly under the key lights [9] Foley could see the shadowy shapes of the crew intermingling and working around the steadfast presence of the Secret Service, the FBI, The CIA and, he thought, there must be some ATF out there also. There was a click in his ear and his ear piece was live. He found he was on the same circuit as the President.

"Mr. President."

"Yes, Mr. Malawi, -- what do I hear from the FBI?"

"Just information." came the quiet reply "We are being tapped from an outside source."

"You mean someone is monitoring us from outside?"

"Yes Sir."

"Do you consider this a threat?"

"Not at this point Sir."

"What does the CIA think, Mr. Tanner?"

"I concur with Mr. Malawi, Mr. President. "

---

9 The large, hot, bright lights suspended on the ceiling grid above every sound stage.

"From what Warden Foley has told me about the Amador group it may be them. Can either of you tell me if it is Luci or Amador?"

"Negative." came the dual reply.

The President was sweating. Somebody reached out from the semi-darkness and patted his brow dry. It reminded Foley of the personal comfort his mother used to give him when he'd scratched a knee. "Two minutes to air time, Mr. President."

"Warden Foley"

"Yes, Mr. President."

"Give Mr. Malawi that unlisted number you have for the Amador group. Mr. Malawi, call him and ask him if he is doing the tap and let me know – Mr. Ohrbach?"

"Yes Mr. President," said the Secretary of State from the control room.

"How many nations will be connected this morning?"

"Biggest audience we've ever had. Last count one hundred twenty-two out of one ninety-two and more expected."

"How can we see them all at the same time? We don't want ruffled feelings."

Ohrbach chuckled. "The director tells me he can split the screens and get up five times that."

Foley's line clicked in his ear and Malawi asked, "Warden Foley?"

"Yes, just a minute." Foley searched through his pockets, found a slip of paper and read the number to Malawi.

"One Minute."

A final scurry of bodies from the set and Foley and the President shrugged at each other. They stared into a bank of stationary cameras, each with its human appendage attached by numerous cables. Beyond them, out of the range of lights roving mobiles would supply the shot maker with numerous extra candid choices from odd angles to feed to the closed circuit audience.

* * * * *

"Mr. A, Mr. Malawi, claiming to be FBI is on the line. There is a country named Malawi in my files and there is also an FBI Special Agent."

"Cray, Give me the info on the FBI Malawi."

46

Paul Ashton listened with astonishment. "Tula Malawi. Now FBI Special Agent, Washington D.C. African American, six foot seven inches, ex-NBA Laker's basketball star. Graduate Chicago Law School, top five percent. Mother was Nigerian, maiden surname Okomolu. There is more."

"Cray, that's enough. Open the circuit. – Mr. Malawi?"

"Yes, Special Agent Malawi, is this Mr. Amador?"

"Yes Mr. Malawi, what was your mother's maiden name?"

"I'm with the President, Mr. Amador, I don't have time to play games."

"Good day, Mr. Malawi" Jason said without hesitation and hit the kill button.

Peter grinned. Paul said, "Damn Jason, suppose it was the real Malawi?"

"If it was he'll call back and this time he'll be more cooperative..."

"Mr. A, Mr. Malawi is on the line."

Jason winked at Paul. "Cray, open the circuit."

"Amador here."

"Okomolu."

Jason smiled. "A fine old Nigerian name Mr. Malawi, what can I do for you."

"We are being monitored. Is it you?"

"Yes. We are monitoring your incoming signals and currently there is no one else."

Malawi's voice came quietly, tiredly into the room. "Tanner and I were hoping it was you. I must hurry this but I need one answer before I ring off. The subset of sound you described to Foley, can we recognize it?"

"A low pitched hum almost inaudible to the human ear. Tell your technician to go to Delta level and you'll pick up the pattern. It is only significant because it was present all the time Luci was there."

"Thanks, may I call you again?"

"Yes and good luck."

"I hope we don't need it."

* * * * *

"Thirty seconds. Lights down on the set, give me quiet on the set please." came the Director's terse command.

"Mr. President?"

"Yes Mr. Malawi."

"It is the Amador group that is monitoring us."

"That is a great relief to me, thank you."

"Quiet PULEASE, Mr. President. Monitors up, please and make 'em live but no sound." In an instant the wall of monitors blinked on as one, looking like a fly with a hundred different eyes. It was a collage of angry, confused, frightened people hung on that wall.

"All mikes up half, Mr. Secretary's mike to full, Five, four, three, two, go Mr. Secretary..." Most of the faces on the monitors were not immediately aware that they were being projected into the United States studio but as their crews informed them that they were 'on' they began to disassemble and, as the Secretary began to speak from the control room they more or less came to order. But as he brought them up to date on the events leading to the conference, there were grunts of scorn and anger and muffled cries of "Great Satan", "Yankee Imperialism" and "Capitalist Pigs."

As the din became louder even with the mikes at half the Secretary held up his hand and stopped speaking. It took a full minute for the crowd to come to order and even then there were two or three loud dissenters.

He resumed. "My Commander-in-Chief, The President of the United States is standing by to continue this discussion with you but I must remind you that this situation is most serious and if you continue your outbursts I will simply cut your sound off. – Mr. President."

The lights came up and the camera displayed the grave visage of the President of the United States. He took up where the Secretary had stopped. Foley watched their faces as emotions swept across the wall in waves. Anger, frustration, defiance, fright -- all backed with a tinge of terror. And when the President ended his explanation with..."and if Luci should appear at this gathering we will abruptly end our transmission." the wall of faces fell silent.

The President introduced Foley. As the spots picked him up he told his story simply and quickly then the President opened the forum to questions. Bedlam followed as the President attempted to field the questions one by one but it became an impossible task, everyone was shouting at once.

The President signaled Ohrbach in the control room and all incoming sound was squelched. "Mr. Secretary, take over and take questions for about fifteen minutes, then boil those down to the most important half dozen and I'll answer them. In the meantime make damned sure we're not on camera."

* * * * *

"Commander? Is all in readiness?"

"Yes, Number One."

"Good. All electronic movement is being monitored minutely from this end so I will not contact you for sometime."

"I understand."

"How is the astral programming progressing?"

"It is ready for testing."

"The girl is the best subject. Test it on her."

"Yes Sir."

"One out."

* * * * *

"Cray, is anyone else monitoring that meeting?"

"Negative, Mr. A."

"Interesting. They are sophisticated enough to recognize the tap but not who. Cray, the order still stands -- I want to know the instant there is any fluctuation in their incoming signal."

Jason, Christa, Peter and Paul Ashton were literally watching the wall in the CPU. On one wall Jason had duplicated the wall of faces facing The President and Jed Foley in Washington D.C. On the other he had the stage set designed for Foley and The President which was now dark and vacant.

Ashton sat back in his chair and gazed in open mouthed wonder at the spectacle on the wall. Jason had cut the sound for convenience but everything that was registering electronically in Washington was projected on that wall. "Jason, I've seen some amazing computers, but Cray makes them all look like tinker toys. How have you done this?"

Jason drew a self indulgent breath. "She is rather special isn't she?"

"You're being modest. How'd she come about?"

"I know most of the Cognitive Scientists at Cray from my school days. When I got ready to build Cray they took the latest tweaks they had on the drawing board and I added a few of my own and they built the core for me. Cray is actually three mainframes which is one reason she is so fast."

"How fast does she actually run?"

"The original cores ran at over eight hundred megahertz."

"Christ! I thought mine was fast at 233" He stopped and looked intently at Jason. "You said the 'original' cores. Is Cray faster even than that?"

Jason smiled. "I'm afraid so."

"What did you do?"

"I uncovered technology. Ordinary technology, even as advanced as Cray's won't allow speeds much above that and even then the heat she created when working full blast was awesome. It boiled down to decreasing distance and thus time. Engineers were stuck in a spiral of piling transistors on top of transistors and cluttering the board with more miles of wire all added to give them more power."

"That's still the way everybody operates -- what's your trick?"

"It's really very simple. I went smaller. I designed a silicon chip within a chip, within a chip which in effect funnels the binary data through ever decreasing space and time."

"The farther she goes, the faster she goes."

"Something like that. As a final kick I replaced aluminum wire with copper wire and optic cable wherever possible, went totally digital and made her voice sensitive."

There was revelation on Paul's face. "She truly works at the speed of light then."

Jason's face was smug. "She calculates at somewhere around 15 gigahertz [10] on three different levels. What is really awesome is when I turn all three levels loose on the same problem. A friend at the University of Memphis helped me design a different approach to time

---

[10] A gigahertz is one billion. This would be 15 billion operations a second.

sharing so that Cray can allocate portions of the work at these hyper-speeds."

"Why don't you patent this technology and turn it loose on the world? It would make you a billionaire overnight."

Jason grunted. "I've got all the money I can ever use and I don't want the scrutiny."

"Look, Jason," Ashton pointed to the screen. "Turn up the sound. Looks like something is happening."

* * * * *

"Mr. President?"

"Yes, Mr. Ohrbach."

"To bring you up to speed the worldwide toll from the bacteria is now over one thousand confirmed cases with half that many still waiting diagnoses. The fax and phone lines are jammed and even the Net is getting overloaded."

The chief Executive sat in the dim sanctuary of his private chambers, shoulders hunched against the news; he ran his hand across his face. "How're we doing with the crowd?"

"I think we've got this down to manageable proportions now. These are temporary elections for this meeting only. The Muslim Nations have elected the Ayatollah Hashemi of Iran to speak for them; the European Nations including the U.K. have elected Sir. Robert Brownlee, Secretary General of the United Nations to represent them; China, Japan and the rest of the Orientals have united under Cho Lin Hui, newly elected President of Taiwan, although North Korea is threatening nuclear war. General Armando S. Niho, Prefect of Brazil, will represent South America."

"How we doing for translators?"

"We're in luck. All these gentlemen speak fluent English."

The President grimaced. "At least we don't have to find someone who speaks Farsi on such short notice. The Ayatollah will be the usual pain in the ass and Niho will preen and pomp but I should be able to get some sense out of the rest. Let's go back to the set. Tell Foley I want him there also -- he's the only one of us who has seen this bastard."

* * * * *

The room was calmer now. Only five portraits studded the wall and though their philosophies could not have been more divergent those individuals looked ready to attack the problem with serious purpose.

The President seated himself, flicked his mike and asked the director to raise the lights. As he came into view the room hushed. He raised his hand and spoke. "Gentlemen. Thank you for assuming this responsibility so we can progress. Last minute reports give us more than one thousand bodies throughout the world and more on the way. Each of you have been hit by the bacteria so you know it as fact. Each of you, as well as the people you represent have been given a transcript, photos and a copy of the Luci tape. They are identical and embody everything we know about this monster. There are only twenty-five people who have seen him so far. One of them is dead and one of them sits next to me. I'll save you time and him agony by telling you that he can add nothing to what you already have in your possession.

"Now the money..."

A clamor from the moving portraits on the wall shut him off. The Englishmen's voice was loudest. "Mr. President, this is absurd. This, this hi-tech moving picture is holding us all hostage. We don't know who he represents, where he comes from or even what he ultimately intends to do with us. I say that we defy him, defy him and let him do his most dastardly. We in the U.K. do not intend to knuckle in to him."

As he paused the Ayatollah struck his most dramatic TV pose. "It is our consensus that this is the infidel's twenty-first century crusade. Hark ye well that we are not the easy marks we were in the ninth century. Today we have the most sophisticate of weapons..."

Niho slammed his fist on the table in front of him drawing temporary attention. "Shut up you stupid rag head, you must not be able to read your own news reports. They are screaming about Muslims drowning in their own blood. If you read any outside news you'd know that it is hitting the world the same way. Now is a good time to be poverty stricken, they are the ones who are getting hit the least. I just lost my Chief of Staff..."

Conveniently a wild wave of his hand knocked the bank of mikes off his table and Niho went on in silence. In the void President Cho

Lin Hui, the people's choice from Taiwan said. "News travels slowly in the back countries so we do not know about the small villages. But our toll in the cities is beginning to reach epidemic proportions. On the island we are hardest hit in the foreign technicians. Have you no idea what we are facing?"

The President said. "It is a short lived bacteria. We don't know how it is delivered except probably through the air. We do not know why it picks any particular victim. We know no antidote nor cure. The onslaught is vicious and the end is terminal. The only thing we know for certain is that this 'devil' has vowed to bleed the earth dry of its resources and leave us all clutching at each other's throats..."

"You referred to the money..." Hui interrupted.

"...Yes. In order to give the world some breathing time we have decided that we, the United States, will provide fifty   billion of the original one hundred billion demand. The rest of you..." and the President spread his arms wide, "...will have to come up with the rest." He looked hard at the Ayatollah. "Now is the time for you and your OPEC friends to put some of that oil money to good use for a change. That aside, the purpose of this meeting is to inform you of our decision and give you time to make the funds available..."

The Brit broke in. "I have word here that the Swiss are staking their neutrality again and are dissenting. They say they will not get involved..."

The President exploded. "You tell those goddamned mountain climbers they'd better cough up or I personally will set troops a yard apart around that country and starve them into history. This is not a regional spat we have here, Gentlemen, this -- this 'thing' is ringing the world's death knell and any one who expects to be around when it stops tolling had best ante up. Mr. Ohrbach will give each of you a phone number which will be manned twenty four hours a day." He looked at his watch. "We have exactly thirty hours and twelve minutes until Luci's deadline. Get on with it and keep in touch," and with that he abruptly left his seat and stepped away from the light.

* * * * *

CHAPTER 11
SUNDAY 13:00 EST, 18:00 GMT, - 26.5 Hours

"Mr. 'A', Special Agent Malawi is on the secured line."

"Cray, have you confirmed his voice pattern?"

"Affirmative."

"Cray, open the line. Agent Malawi?"

"Yes, Mr. Amador?"

"Here -- what's happening?"

"The President has given me instructions to keep you informed. But first, did you detect anyone else monitoring our signal at the White House?"

"No. What news are you getting from around the world since the President signed off?"

Malawi cleared his throat. "Interpol, MI-5, the KGB and the Mossad have all agreed to share information using us as the central clearing house. The known death toll is now above thirteen hundred."

"That confirms our figures. Do you think the rest will share information?"

"No more than absolutely necessary."

"How about the other side of the fence?"

"They don't trust us enough to talk to us. The North Koreans are threatening to nuke us, the Fundamentalists are waving the terrorist sword, and a good part of the world thinks the rich Jews ought to donate the whole one hundred billion. Of course all of them would just be pleased as hell if Luci wiped the Western World off the map."

"How about on our own side?"

Malawi chuckled. "The situation is not acute enough yet to cancel out old scores although Tanner and I are working quite well together."

Christa said, "Jason, ask him where the meeting is going to be held?"

"I heard that. Mr. Amador are we not alone?"

"No, three of my associates are here in the CPU. I vouch for them. We are all working on this problem. The voice you heard is that of Christa Bonamay. She is my right hand on the computer."

"Who else is present? Names please."

"Paul Ashton, Warden of Folsom Prison and my pilot/engineer Peter Chaney."

"Thank you. We are setting up in the main chamber of the United Nations Building, Ms. Bonamay. It is designed for large meetings and the full assembly will be at this one. Security is a normal thing there and we have full electronics. Will you be monitoring again?"

"Yes," Christa answered. "And if you will leave us an open band it will save me taking time to hack into the main system there."

Malawi laughed. "You make it sound like a walk in the park. This is one of the most secure systems in the world."

"Every room has a door," Christa rejoined. "as you discovered at the meeting this morning."

"Touché. After this is all over I'd be very interested in seeing how you accomplish that." said Malawi.

"Can you tell us what scanners you'll be using?" Jason asked.

"Is this line scrambled?"

"Yes."

"Every electronic detector available. Infrared, UV, Motion, Electro-magnetic, sound monitors, seismic registers – you name it, and of course, old fashioned metal detectors. There's even a new Positron Scanner that's strictly experimental that we will have set up. We'll be lucky to hear anything without static. Physically the guard will be quadrupled and every square inch will be observed by cameras. Tanner and I will be in the control room until trouble comes."

"Will the President be there?"

"Not in person. He will be in a secure place connected by electronics."

Christa indicated a flashing red light to Jason. It was the quiet signal Cray used when she wanted to speak and there was a conversation already in progress. "I've got some material coming in that I have to attend to. Can you leave a secure phone number with Christa?"

"Yes."

"Thanks," Jason said and turned away. He slipped a set of headphones on, plugged himself in and said "Cray, what do you have for me?"

Cray said without hesitation. "The magazines came from two publishing houses. No staff overlap and they were printed by two different printers in two different cities."

"Damn. No correlation there."

* * * * *

CHAPTER 12
SUNDAY 14:00 EST, 19:00 GMT, - 25.5 hours

"Cray."

"Yes, Mr. A?"

"Get Mr. Chaney a first class round trip ticket, return open, on the Olympic Airlines flight leaving nearest after midnight tonight from JFK to Athens, Greece, Hellinikon Airport."

Peter raised his eyebrow at Jason. "A short, but interesting vacation for you in Athens, Peter. I want you to go to the spot where Luci's transmission originated and see what you can find. Act on what you find. I have a feeling that Luci's base is over there."

He called to Paul and Christa. "Please come here and gather round." They both pulled up chairs and waited. "Christa, you've got Cray tuned in where Paul and I can handle it if you will give him some quick instruction. I want you to take the chopper and drop Peter off at JFK by midnight then make your way back to Baltimore and in the morning go into that restaurant and see what you can find out. I'll tell you where to stay before you leave. Paul, you will stay here with me and we'll run the store. Any questions? -- No? good. Christa, give Paul a short tutorial. Peter, come with me to the lab." He turned, picked up his cane and made off for the far door to the lab. Paul beamed a mischievous smile at Christa and a frowning Peter followed Jason.

* * * * *

Jason indicated a chair. "Sit down, Peter while I show you my latest life saver -- yours." Jason opened a glass case standing on the lab table and extracted something with a fine pair of tweezers. It was flesh colored and about half the diameter of a pencil eraser.

"This is SAM. It stands for 'Static Assimilator Module'. I've been working on this little idea for some years and luckily, it is ready to use."

"What's it do, Mr. A?"

"A number of things Peter." Jason slid a loupe onto his eye and inspected the chip against a bench light. He laughed. "SAM is a formidable weapon and a cloaking device all in one. First let me say that static electricity is present in almost every condition we encounter in our lives. SAM loves static electricity. He loves it so much he spends his entire life accumulating and compressing it. At the same time I've taught him how to kick up the amperage. He goes about this silently while you do other things but when you tap SAM three times he releases the electrical energy he has stored up. First he forms an 'envelope', a force field, if you will, around you. This cocoon of pure energy is lethal..."

"How?" Peter said, puzzled.

"...Don't get ahead of me. Within fifteen seconds of the third tap you must make a choice. If you do nothing the cocoon will implode on you and incinerate you. However, if you point your arm like a lightening rod with your finger extended at a target SAM will discharge everything he's built up in a bolt similar to Luci's and you can send it nearly a hundred yards."

"Is this how Luci does it?"

"No. Here the subject must create the electricity. Luci is not solid so does not create. Luci draws his power from those around him."

Peter looked closely at the chip. "How do I carry this thing, Mr. A?"

Jason grinned. "I'm going to glue it to your scalp..." Peter winced. "...just be careful if you wear a hat. Bend your head over here and hold still."

As Jason began to slide the chip down through Peter's dark hair to the scalp he went on. "You should feel a slight tingling while inside the cocoon – nothing unpleasant. Another plus is that during the fifteen second countdown the cocoon will stop incoming fire – probably up to fifty caliber."

"Will it stop Luci's bolt?"

Jason was thoughtful as he pressed the chip tight against Peter's scalp. "Unknown Peter, but I don't think so. Putting that kind of voltage and amps onto the shield might implode it. A glancing blow maybe, but a direct one, no. If you meet Luci you'll just have to use your wits."

"How often can I fire it?"

"Three or four minutes if you were rock'n rolling on a nylon rug, probably thirty minutes under normal conditions"

"Kind of a one-shot deal," Peter mused.

"Not totally so," Jason rejoined. "Subsequent bolts wouldn't burn through steel but they would certainly give a man an awful shock."

* * * * *

Paul Ashton watched Jason and Peter as they disappeared through the door to the lab. He turned to Christa and before she could speak he said. "May I ask you a personal question?"

She looked at him with wary blue eyes. "Yes."

He stepped close to her, brown eyes staring intently into hers. His question was direct. "Do you have anything going with Peter?"

She chuckled. "Peter and I are good friends and we work together -- that's the only thing we have 'going'".

"He'd like more," Paul said

"Well, most men would..." she said matter-of-factly and decided to meet the issue head on. She speared him with those brilliant blue eyes. "...and you?"

Ashton blushed. The intensity of her gaze was enough to make his blood pound. Twice divorced, ten years a bachelor, immersed in his work he thought he had muzzled and buried the juices that came flooding back through his body and into his consciousness from the first moment he had seen this tawny beauty. He had labored through the afternoon talking with Foley, the other Wardens and correlating information for Jason and carrying on light conversations with the other three including Christa but he could not take his eyes and his mind off her. He blushed again. At this moment he was a fourteen year old schoolboy ready to lay down his life for his lady love. In his fantasy he swept her into his arms. "I, ah, I must confess, Christa, I'm smitten. I have never met anyone quite like you. I am drawn to you like – like a bee to honey..."

She pushed him gently away. "Whoa, Paul, this is going too fast. We just met this afternoon..."

Paul held her at arms length and rushed on. "...I know but if Luci has his way we won't be around much longer and I don't have any time to waste -- at least I want you to know how I feel." He finished lamely.

She took him by the shirtsleeve and led him like a first grader over to the main console. "At this moment we don't have time for personal feelings," she said gently. "In order to help Jason while I'm gone you've got an awful lot to learn about Cray."

Paul laughed nervously, "Forgive me. For the past ten years I've drowned myself in my work. I've totally ignored the softer side of life so I'm badly out of practice in the ways of saying personal things to a woman. All I know is that I haven't had these feelings, these yearnings for so long that its like the first time..." She started to interrupt but he shook his head and held up his hand. "and sometimes its so obvious it embarrasses me to face you and I don't know whether to stand up or sit down."

She sought to ease the tension. "Paul, if its been that long since you've been with a woman then that's the simple explanation. You've denied mother nature too long."

Now Paul was totally embarrassed. His face turned a deeper shade of red. He ran his hand through almost nonexistent hair. He stammered, "It, its not only that Christa. I hadn't realized how much I'd missed the softness, the gentleness, the warmth of a woman's presence. I thought I'd learned to put all that away from me -- but this evening being around you has been a blazing revelation. You've brought it all back with a vengeance and here I am at the age of fifty stammering and stumbling over my feet and my words like a pimply schoolboy." He turned away from her. "Jesus! what an idiot I am."

"Paul," she said quietly. "Another place, another time, maybe something might come of this. But right now we literally have the weight of the world hanging on us. I really think Jason is the only one who can figure this 'thing' out and he needs all the help we can give him. Please," she said tugging at his elbow, "help me help him."

Paul's voice was husky, his face haggard. "Of course I will, -- you know I will, but I can't help how I feel about you."

* * * * *

Jason inspected the glue on Peter's scalp and nodded his head. "You can comb your hair back now and no one will ever know you are wearing SAM." Peter pulled his comb from his pocket and Jason walked over to a long, gleaming bench and opened a drawer. "I have a couple of other tricks that will make it easier for you to get back." He held up assorted parts then quickly snapped them together. In a moment he held up a gleaming white pistol and tossed it to Peter. Peter was astounded at its lightness.

"It's plastic!" he snorted.

"Ah yes but what plastic. A wedding of Polyvinyl chloride and Polymers. Forty caliber, accurate to one hundred yards and fires a Teflon coated, gas propelled high explosive round that will decapitate a man." He tossed a clip of stubby ammunition to Peter.

Peter caught the clip. "How do I get it there, Mr. A?"

"Break the pistol into two parts and put one in each pocket. The rounds and the pistol will go through airport X-Ray."

Peter looked at the packet of ammunition. "There's only ten rounds here."

Jason shrugged. "You're from the throwaway generation. After ten rounds heave it in the river -- it'll be pretty well burnt up inside anyway. Now here's number two."

He indicated two slender, black plastic boxes about the size of a TV remote on the table in front of Peter. Peter picked one up. Jason spoke, "The second one is for Christa. You can explain it to her on the flight. That is small enough to fit in your shirt pocket and has seven distinct operating modes. One -- Through GPS[11] it'll give your position anywhere in the world to within three meters -- what country, what city, and your address. Two – drop coordinates into it and it'll tell you which direction and how far to get there; three – it'll sweep for electronic bugs of any kind; four – its an infrared scanner; five – it's a powerful cellular phone good any where a satellite's overhead; six -- a digital camera connected to Cray; seven – it all masquerades as a small Shortwave and AM/FM radio which actually works – which means that it too will go through the airport. All the other controls are underneath the radio."

Peter had been examining the radio while Jason talked. It took a moment to decipher Jason's instructions on manipulating the dials and switches then he nodded his head and dropped the box into his

---

11 Global Positioning System.

shirt pocket. Jason put his hand on Peter's shoulder. His gray-green eyes searching Peter's dark ones. His voice was troubled. "Peter, At the moment Luci is beyond my ken. I don't know what I'm sending you into nor what you'll be up against -- we simply haven't enough information. Until we do we'll feed you whatever info we get that will help combat this thing. Nobody's better prepared to handle this than you are so you'll just have to wing it."

He turned to the table and began to put assorted small tools back into their proper place in the drawer before him. "Christa and I do want you to come back to us, you know." he added dryly.

Peter grinned. "Mr. A, I haven't had any action in a long time and I'd rather go over there than sit here in the CPU. But you've got to promise me something."

"Eh, what's that?" Jason looked up from the drawer.

"I like the Warden, so don't get me wrong, but I think he's got a crush on Christa and I don't trust him, so watch him for me will you?"

Jason chuckled and gazed off into the distance. "Paul has batched it for ten years. I had no idea he might be smitten." He looked at Peter. "In case you hadn't noticed, Christa is a beautiful woman. So you think Paul might move on her, eh?"

Peter winced. "We-el, maybe."

"She's been celibate for a long time Peter. Maybe Paul has set the juices flowing."

Peter bristled. "That fat, bald headed old man. What could she possibly see in him?"

Jason smiled. "History is full of bald headed lovers, Peter. Besides, second to the onset of puberty, the fifties flush is the most intense sexual period a man goes through. At that age the divorce rate soars, philandering is a national pastime and many, many babies are born to over-fifty fathers." He closed the drawer while trying to hide his amusement. "However, with this new found information I'll keep an eye on him. Let's go back to the CPU. I've got some last minute words for both of you."

* * * * *

Jason stood in front of one of the large screens on the wall. It looked like a racing tote board with figures advancing as he spoke.

Cray had alphabetically listed the country's of the world and was keeping a tally of the deaths each was reporting that they attributed to the bacteria. The total stood at over four thousand and advanced steadily.

The other three sat and stood around in various positions. He toyed with his gold headed cane. "Almost always when you have an "unknown" phenomena there are laws of physics that apply and partially explain the aberration. So far, we haven't got enough information on Luci to form more than primitive opinions on his makeup, his MO[12] nor, most importantly how to combat him. We do know that he has now shown himself at least five separate times throughout the world, including the original at the Warden's meeting- and has personally killed four people..."

"Always the same way?" Paul asked.

"Yes, the same bolt and the killing hand appears to always be the right." He produced a small, clear plastic box from his pocket and handed it to Christa. It was like a very small egg crate and visible inside were two capsules about the shape and size of a large amber jelly bean. "Here are some very special locators so that I can tell where you are at any given moment. I want you and Peter to each take one."

"How about the locator I've got in my shirt pocket?" Peter asked.

"These can't be lost, taken away from you or misplaced. It is absolutely imperative that we know where you both are at any given time when you leave here. Take it with any liquid and swallow it whole. As the soft gel dissolves[13] an adhesive evolves that sticks to human tissue. It'll stay inside for about two weeks then the adhesive will dissolve off and the capsule will come on through."

"What's inside the gel?" Christa asked as she passed the little box to Peter.

"A battery and a simple micro chip that puts out a constant signal to Cray through the G.P.S. Peter, is the chopper ready to go?"

"Yes."

---

12 Modus Operandi (L) Manner of working or operating.

13Made from poly(lactic -- coglycolic) acid used in sutures that dissolve in the body.

"Good, then you youngsters need to be ready to fly in the next thirty minutes." He looked at his watch, picked up a remote and flicked it. A large map of the eastern seaboard popped up on one screen. He pointed to it and said to Christa. "It's about 835 air miles up to JFK. I want you to lay a course over the mountains here. Split Raleigh and Fayetteville -- lock onto the beacon at Nags Head and on out into the Atlantic beyond the Banks – that's roughly four hundred fifty miles and a fraction over halfway. Continue offshore for about a hundred miles and then almost straight north over the water and pick up the beacon into JFK. You should be far enough away from the shore to not disturb the Norfolk/North Hampton security system but Cray has posted a flight plan already so just acknowledge and fly on if someone hails you. You'll be coming in to JFK over the marshes of Grassy Bay. If you stay low from the Banks up you shouldn't have to worry about traffic. Over the water you should be able to ram it up around 150 knots until the last fifty miles. Then they'll pick you up on radar and the traffic will get heavy. With the flight plan no one will be surprised by your arrival. They'll talk you into the right landing pad. Any questions?"

She shook her head. "Yes, where do I refuel?"

Jason continued. "In a moment. It's nearly 4 p.m. now so that should put you in JFK a little before ten to drop Peter off and then get you back to Baltimore before midnight."

Christa asked. "Where do I stay in Baltimore?"

"Same place you refuel. Believe it or not the Strasbourg, on Light Street a short walk from Phillip's, has a rooftop chopper pad and fuel. They are expecting you when you get there and will maintenance you while you sleep. Peter, I've not made a reservation in Athens for you because I don't think you'll be there long enough to use it, but if that changes I can always call in a favor at the Palacio. The owner there is an old school chum. Christa, you'll need to pull ten thousand from the safe, – keep one and give Peter nine."

He fussed with his cane for a moment while looking fondly at each of them in turn. His voice was grave. "I can't express to you how worried I am about your welfare out there. All I can do is tell you to exercise the utmost caution." He looked directly at Peter. "I want no dead heroes here Peter, what I want is you two young people back here in one piece, -- understood?"

64

Peter grinned and nodded his head.

* * * * *

CHAPTER 13
SUNDAY 16:30 est., 21:50 GMT,   -- 23 HOURS

On the screen Paul Ashton watched the chopper fade to a tiny speck, then to a pinpoint then to nothing in the northeastern sky. He punched the maximum resolution button again just to make sure. Cray said, "I am already at maximum."

"I know it Cray, I just hate to see her go out there by herself."

"Miss Christa is not by herself." Cray corrected.

"Yes, and that's half the problem," Ashton snorted. "I would feel one whole hell of a lot better if she were."

* * * * *

The chopper sped away from the late afternoon sun on its hurried way across the mountains. Their dragonfly shadow flitted ahead for a few minutes until the fading sun drew its last gossamer pink from the edges of the clouds and their reflection dived into the gloom. Soft blue night settled over the landscape and clothed the bird in a velvet sky topped with rhinestone stars. Christa was always deeply moved during night flying. Inside the cockpit the muted green panel lights gave just enough light to read the dials and follow the needles. Through the window the horizon was sprinkled with sparkling, sentient diamonds. It was like being enfolded in a warm, black cave suspended above the occasional twinkling lights. Of the world but not part of it. You could see, you could watch, sometimes you thought you could almost hear the mélange below. You were part of the world, but not touched by the worries, the cares and the woes of the ants below.

She reveled in it, but she also knew the hypnotic drone of the blades merged with the high pitched whine of the wind would bring on torpor very shortly so she leveled off at five thousand feet, kicked in the autopilot[14] and engaged the HUMS[15]. Rotor speed increased and airspeed went up and leveled off at a blistering 150[16] knots.

---

14 Autopilot: A mechanical device that will fly the helicopter on the level and through an ILS

Peter had furtively watched her but left her to her concentration since they had left the pad. She turned to him and he smiled. "You did all that like an old hand, Christa.

She smiled in the green glow, "The best teacher in the world couldn't have anything but a good student."

"You like to fly?"

"Yes, it rejuvenates me."

Peter watched the velvet ground slide along beneath them and disappear into the velvet sky. Every star in the sky a point of light glistening like wet diamonds. A more romantic spot he couldn't imagine. A more beautiful companion he could not wish. "Loving and flying are very much alike," he said.

"I find far more solace in flying this chopper than I ever did in the so-called love given me by my tricks back in Jamaica." She sighed and went on checking the green instruments.

Peter chuckled, his voice seeming far away through her headset. "There's not much solace in snuggling up to a vibrating chopper and it certainly can't give you any warmth."

"Much safer though, less hurt, less pain."

Peter turned toward her, trying to make eye contact. She deliberately turned away from him and gazed out the window. "You are really wasting the best part of your life, you know." he said.

She did not answer. He waited a few moments then decided to change the subject. "Are you attracted to Ashton?"

She watched the last tinges of red fade from the high eastern sky. The sky and the earth fused into a black cave and the droning chopper chewed its way into it. Was she attracted to stodgy Paul Ashton, pudgy, middle-aged and balding. He represented a part of life foreign to her. Stability, reliability, permanence. How important is respectability, she wondered. Will that package take the place of the

---

(instrument landing system) approach and level off at 80 feet and hover.

15 Health and Usage Monitoring Systems: Sophisticated electronic computer devices that, by continually monitoring parameters such as temperature, pressure, and vibration level, can determine the condition, or health, of mechanical components. In helicopters, the engines, main gearbox, drive train, and tail rotor gearbox are typical candidates for HUMS.

16 172 mph

pulse quickening tingle in the look, the intimacy in touch and word, the promise of ecstasy to be fulfilled that happened every time she brushed against Peter. How much of yourself do you give up for those rewards -- how long do they last?

"I'm comfortable with him," she said.

"Couldn't you be more comfortable and have more fun with someone like myself -- someone more your age and style?"

"Ah yes, Peter. But if I give into your charms I'm a slut not worth worrying about and if I don't I'm just an old whore. Either way I come off a loser."

Peter winced. "Man, have I got that coming. Are you ever going to let me off the hook on those stupid remarks."[17]

She smiled inside her helmet but turned her head away from him so he could not see. "Perhaps, when I feel you've suffered enough." Her words were blunt but the tone was softer than the words and left him some hope.

He lost himself in the inky blackness of the deep woods below. The swift little chopper hummed its solitary way beneath the canopy of stars like a foraging honey bee on its way back to the hive. Here an occasional light blinked on the ground, then disappeared. There a twisting ribbon of moonlit road snaked through the hills, crossing a glistening river winding in and out of the shadows. Far off against the horizon a lake turned silver than black as the moon cut a path toward them. Much of the beauty of the landscape was lost to him as he gritted his teeth until his jaws hurt and began to probe the depths of his own stupidity. The difference between simply living and being alive is being with someone. Not just anyone, but that special someone who brings the light, writes the music and sings the songs of love. She first fans the flames of lust into an unbearable urge which only she can sate, then blithely casts her spell of permanent love with visions of hearth and home and husbandry. How, he thought, do I keep getting off on the wrong foot, -- no how can I ever get off on the

---

17 Read "The Voodoo Vortex" by this author. Christa had been a teenaged whore in Jamaica long enough to make enough money to come to the US and put herself through school. Since then she has been celibate. In their first interaction together Peter was less than kind in his remarks and only began to see the real Christa after she saved his life.

right foot with this gorgeous creature who drives me insane when I get into the same room with her. I'm sitting here in this chopper with her and I feel this over whelming urge to reach out and grab her, to hold her, to kiss her, to make mad, passionate love to her -- and she's got me down and won't let me off my back. When I'm with another woman I'm simply living but this woman brings me ALIVE. She whips me into a frenzy while she tears at my nerves, slaps my feelings around, rides roughshod over me, spits at my apologies and laughs at my ineptness. I feel like an awkward schoolboy standing in front of the beautiful teacher. God! This girl gives me a hard on and each time leaves me with it....

A slight lurch and a definite swing to port interrupted his reverie. He had not noticed the passing of shadowed land into shadowed sea but now, as the Auto Pilot made the northerly course correction he realized they were far out over the grey, white capped Atlantic. He gazed at the unappealing water and shuddered, he had no wish to go down at sea.

He turned the radio on scan and listened intently for just a few moments to each station as the radio paused. There was nothing mentioned on the regular bands. Evidently the whole commercial world was keeping the catastrophe under wraps and off the airways. Trying to avoid general panic, he thought. It was only when he got into shortwave and the Ham Bands that he caught smatterings but most of it was delivered by non-professionals and mostly panic laced misinformation.

* * * * *

Christa double checked the autopilot and glanced sideways at him. "How much time you figure we've got to New York?" He asked.

"Roughly two and a half hours. I can't sleep, why don't you?"

"Good idea in a minute," he answered, turning the radio off "but first I need to explain this little box Jason asked me to give to you."

To Peter it took all to short a time to give Christa the operating instructions on Jason's invention. She caught the information the first time, asked a couple of quick questions and their intimacy was over. He grumbled something about being sleepy and reluctantly slouched back in his seat.

* * * * *

It seemed but a few moments when he heard Christa hailing JFK Control and getting their laconic answer. He looked out his window to see the mid evening lights of Asbury Park twinkling and sliding past the port side of the chopper. Just ahead was Sandy Hook and across the channel the lighted gash that was Coney Island.

Quiet words from JFK prompted Christa to cut her airspeed and correct to starboard and they beat their way to the right of the crowds at Rockaway Beach and out over Jamaica Bay. Light fog lay above the water below them and was whipped into a frenzy by their down draft. Further instructions from the control tower prompted Christa to turn to the West in a wide circle out of the path of larger air traffic. As they passed over it Peter pointed to the huge causeway that spanned the bay from Holland to South 92nd St. "That's Cross Bay Boulevard. We've only got about five minutes now before we get to JFK." He pointed toward the near horizon. "In Fact you can see the lights and the runways through the mist over there."

The control tower came in again with crisp instructions and in minutes Christa had reached her designated pad and let the bird settle gently to the tarmac.

While they still had headphones Peter asked. "Have you talked to Jason?"

"Yes, I told him we were landing."

"Did he mention what's happening?"

Christa shrugged. "The body count is now almost double what it was when we left L"Aerie and plans are progressing for the meeting tomorrow."

Peter whistled softly. "I found very little about it on the radio earlier"

"I know, I tried also and it looks like they've put a general clamp on all the commercial news."

Peter collected his gear and dropped to the ground. Christa finished her flight log and joined him. Ground crew appeared, she signed papers authorizing fuel and they walked into the small plane terminal. He stood quietly among the crowd while she signed in, and out, then rejoined him. .

He waved his tickets at a long hallway that merged into a larger part of the building. He turned to Christa, speaking quietly, trying not to show his concern. "It looks like that's my direction and my plane leaves almost as soon as I walk on it."

Christa looked at him, looked away, looked back at him again. A tremor passed through her but her voice was steady. "Peter, be careful – please."

Peter's light hearted chuckle echoed in the now empty hallway. "You mean the ice maiden cares about old Peter."

She snorted. "You know I care about you and I don't want you killed nor even hurt. This monster is deadly and we have no experience with it. "

He turned serious. "I know. Keep me posted with anything you learn that will help me cope with this character. In the meantime give me a big hug and I'll be on my way." Without waiting for a reply he scooped her into his arms and pulled her to him. For the longest of moments he buried his face in her fragrant hair and breathed deeply. For those same couple of seconds she responded with soft breasts and supple form molded firmly against him. Then she kissed him gently on the cheek, stepped away and broke the spell.

"Spoilsport"

She blushed and said. "You've got more important things on your mind and I want you thinking clearly -- not about me."

"That is an impossibility but I suppose I must try. I'll see you soon," and with that he turned and made long strides down the hall away from her.

* * * * *

CHAPTER 14
SUNDAY 23:30 EST; 04:30 GMT MONDAY;   -- 16 HOURS.

Peter had an anxious moment stepping through the metal detector but Jason was right. Polymer does not ring the bell and so he picked up his luggage on the other side of the barrier and walked down the long hall to the waiting room.

The plane was nearly boarded so it was the work of only a few minutes to get on board, in his seat and order a drink from the attractive flight attendant who favored him with a seductive smile.

As the ship began to roll off the loading pad and strike out to the runway he opened a copy of the evening's playbill, noted the five movies available and chose #3, a Cheech and Chong rerun. If it got too boring he wouldn't mind switching it off and in the meantime it might be funny. As the ship began its run down the tarmac he donned the earphones and settled back.

* * * * *

In the darkness of an office, a fax machine began its muted acceptance. The surrounding darkness is so velvety black as to seem underground. The air is sepulcher, cool; there's no sound but the slight whir of the machine. The cryptic message reads: "PC. Olympic Flight #1145, Seat 4A, Line 3."

* * * * *

It was the first time Christa had landed on such a restricted space and she wiped a fine bead of sweat from her upper lip when she finally dropped the little bird gently onto the pad on top of the Strasbourg in Baltimore. Air traffic had been light but she quickly discovered flying in and around tall buildings was not her want and she heaved a long sigh of relief when the little chopper settled in and the rotors began to sag.

A tall youngster in overalls appeared at the door. His face lit up when he saw her. "That was a nice landing Ma'am, can I help you down?"

She looked into the friendly face, smiled her sweetest and noted that his name tag said 'Calvin' as she accepted his outstretched hand. "You must be Miss. Bonnie Mae. I've been expecting you. This is some bird you got here Ma'am, " he said letting his eyes roam lovingly over the Agusta as Christa stepped down to the pad. "I'm gonna have fun rubbing this one down."

"Do you 'rub all of them down'?"

His smile almost obscured his nose. "You bet, that's my job. I fill 'em up, check 'em out and make sure they shine when you get ready to leave."

Christa pulled a fifty dollar bill from her pocket and pressed it into his hand. His eyes bulged momentarily then he started to hand it back. "Geez ma'am, the boss won't let..."

Christa pushed his hand back gently and said, "Calvin that's just between you and me. Just my thanks in advance for the fine job you are going to do taking care of my bird."

"Wow, thanks Miss Bonnie Mae. When you gonna need this bird?"

"Not before late morning, now, where's the elevator so I can check in."

* * * * *

As the jumbo jet lifted into the chill air at thirty five thousand feet and began its eastward trek into the rising sun half of Peter became absorbed in the antics of C&C and the other half lazed back into the seat in anticipation of the hours of flight ahead. His mind turned back to his last glimpse of Christa of the long legs and high bosom. His blood raced and he had to shift his legs to remain comfortable in the deep seat. What was it she had said about being an ex-whore. He couldn't remember exactly but he did know that since the first time he had never thought of her that way. The more he saw of her the dimmer her past became and he wanted her more intensely. But would he still want her if she gave in? For the first time in his life and the first time in a relationship the question of respect came up and it nagged at him. He suddenly realized that it was important to him that he respect her. It never had been before with any woman. They had always flown through his life like a comet, in one corner of his vision and out the other with nothing but a fleeting impression left behind. But this woman was not a throw away line in an off-Broadway stage production to be used to fill a moment and then abandoned without remembrance. No, with Christa, he felt like he was walking through a labyrinth within a maze sheltered in a house of mirrors.

Torpor suddenly switched to alertness. The C&C movie, backgrounded in raucous music and sense pounding, jagged sound effects was in its final throes but suddenly it went to a steady beat with no melody. There was a crack in the air followed by abrupt movement in the aisle to his left. He swiveled in his chair to see the awesome apparition of Luci begin to materialize out of thin air. As Luci took shape he crouched in the aisle. Crouched only because his massive stature left him no space to stand upright. He filled the airplane from top to bottom. He morphed, became smaller, adjusting to the space, but no less lethal as he stretched, stood upright and raised his massive arm to Peter. The arm seemed to emanate from his malevolent smile. Had he shot the bolt then, Peter would have been seared meat but he paused and said, "Chaney you and your master, Mr. "A" are dead."

In times of great stress, training and instinct rule and both served Peter. He ripped off his headphones and flung himself to the aisle floor. A split second later there was a zap like in the old movies. The air roiled with the acrid smell of burning foam and seared flesh. Simultaneously Luci disappeared and Peter's seat burst into flame. He jerked a blanket off a nearby passenger and threw it over the flames as the passenger compartment came alive with moving bodies and terrified screams. An alert flight attendant triggered the oxygen masks and commanded the passengers to don them until the cabin could be purged of the poisonous fumes. The flames went out. Peter pulled an oxygen mask to his face as the First Officer appeared wearing an emergency gas mask and toting a red extinguisher.

His voice was muffled as he demanded of Peter, "What happened here?"

Peter took a deep breath then pulled the mask off his face long enough to say. "I don't know -- my seat just burst into flame.

"Were you smoking?

"I don't smoke – and I don't carry matches."

"Do you mind if I search you?"

Peter opened his mouth to protest but was cut short by a horror stricken scream followed by plaintive cries from other parts of the cabin. "Oh God! My wife is dead," "I can't get my husband to breathe," "My daughter is dead"

The cries came so fast the First Officer could do little more than turn his head from one to the other of the voices. A frenzied woman ploughed her way through the stunned crowd, grabbed the First Officer's arm and screamed, "My son has a hole burned through his head. He's dead, do something!" and collapsed on the floor.

Peter relaxed. The F.O. had other more immediate problems then searching him and had forgotten about him. He offered to help.

"What can you do?" The F.O. gruffly demanded.

"I can carry bodies." Peter responded matter of factly.

"Yes, there will be that problem. Thanks."

As a pair they went to each body and inspected it. There were three adults and five children. The wounds were identical. A neat pencil sized hole drilled from ear to ear and neatly cauterized. A smell of singed flesh but practically no blood and no other damage. He managed to quietly note that each casualty had been watching Channel 3. There were no deaths other than those watching the Cheech & Chong movie. The harried F.O. missed this detail and Peter did not enlighten him. The Captain informed them the cabin air was now safe to breathe so they took off their masks. The First Officer looked at Peter. "I'm not complaining but how was it you just happened to take off your earphones just before this 'incident'."

Peter rubbed his ears. 'My ears are sensitive to pressure. I can't fly very long without taking them off and letting my ears relax."

"Damned lucky, else you'd have been laying down there with the rest of them." He started to turn away up the aisle then turned back to Peter. " Are you a pilot?"

"Yes."

"What class?"

"I earn my living flying a corporate jet and a chopper. Private duty."

The F.O. grinned. "I always dreamed of one of those cushy jobs but I guess I'll always be stuck up in one of these cabins." He stuck out his hand, "Welcome to the family, I'll make sure the flight attendants make you comfortable during the rest of the trip."

"Thanks, are we going to turn around?"

"No, We're just a couple hours off shore so we're not diverting anywhere. If we stopped anywhere other than our destination we'd all be detained for God knows how long so we'll just keep going on to Greece. Are you ticketed beyond Greece?"

"No."

Peter settled into the deep first class seat and reached for the phone. He would have preferred to use his cell phone but the airlines ruled that out so he first asked Cray to scramble the line then left his account of Luci's ravaging foray on the plane so Jason could hear it first thing when he got up. He was as accurate as possible including the tidbit about Channel 3.

He settled back in the seat with a tall double scotch, spent a few well chosen minutes with a Berlitz 'Greek For Travelers' handbook then dropped off to sleep and slept right through to Greece.

* * * * *

The fax machine in the dark room came to life and spit out a terse message. "CB. Strasbourg, Rm. 1104.

* * * * *

CHAPTER 15
Approximately 07:00. EST;12 noon GM; 14:00 Athens, Greece; - 10 Hours.

Christa stretched lazily working the flight kinks out of her muscles. She turned over in the middle of the giant king sized bed and looked at her watch. It's slight tinkle prodded her again. Seven a.m. and time to get up. There was a discreet rap at the door of the suite in the other room.

"Room Service"

"Just a moment"

She popped into a robe, eyeballed the peep hole and opened the door. A young porter grinned at her appreciatively and wheeled the breakfast cart into the room. She indicated where she wanted the cart and watched him helpfully whisk the cloth covering off the table. Gleaming silver and china was set for one and the savory smell of sausage, blackberry crepes and orange juice filled the room.

"May I open the drapes? Ma'am."

"Yes, I think so. Where do I sign?"

He looked back over his shoulder and watched the long dusky legs move inside the robe. "Madam's bill is there on the table. Is there anything else I can do to help you?"

She signed it, shook her head and he left reluctantly.

* * * * *

Jason stumped grumpily into the kitchen. Christa Bonamay supplied many important pieces to the puzzle of his life not the least important of which was her skill in the kitchen. When she was gone he missed her sorely and reflected so as he began to prepare his Spartan breakfast. He heard steps along the outside corridor and turned to see Paul Ashton step into the room.

"Look's like we're batching for awhile," he smiled.

Paul screwed up his face. "I'm afraid I miss her for more than this Jason."

77

Jason stirred his coffee. "It's an affliction Christa leaves us all with. We all see something different in her but we all meet on common ground. We love her -- in different ways of course." He added with a shrug. "Right now I'd like to have some of her cooking."

Paul grinned. "What's the latest news?"

"Christa checked in from Baltimore. She's on her way to the restaurant..."

"Did you talk to her? How is she?" Paul asked anxiously.

"...No, she left her message with Cray. Relax, she's fine. But the world is still going to hell with Luci. The body count is nearing eight thousand. He attacked Peter on his Athens flight last night..."

"My God, is Peter alright?"

"...Yes. Instinct and fast reflexes but Luci added eight other bodies to the list..."

"Good God!"

"...five of them children" Jason added grimly.

"There is some significance here, Paul," he said, sipping his coffee, and went on to repeat Peter's graphic recount of the incident then added, "Luci disappeared the moment Peter jerked his headphones off which was the same moment everyone else was zapped. Definitely has something to do with the sound."

* * * * *

Christa stood in the media strip and waited out the traffic. While waiting, she glanced at the ready light on the small, black box in the palm of her hand. It was pulsing more often now and as she walked across the three lanes toward Phillip's Restaurant which was on the dock down the block the light began to pulse more quickly.

The restaurant was dockside, the first of a stretch of buildings that made up the Inner Harbor complex. Across the short width of the dock was the water occupied by a number of boats of various sizes including a three master across a short waterway. The early morning breeze toyed with its outer jib which was unfurled sufficiently to bear a tattered sign "The Endeavor". The ship was in much better shape than the sign as it gleamed of brass and varnish.

The restaurant did not serve breakfast so the front door was closed as she expected. She shaded her eyes and looked through the window. There was light in the background. A swinging door at the very back opened and flipped light across the room as someone walked through it. The inside of the restaurant was filled with white covered tables, booths along the walls she could see, a long gleaming mahogany bar, and tasteful chandeliers as yet unlit. On her right side the windows overlooked the inner harbor and would be the best seats in the house.

She walked left around the building up the sidewalk into the short alley at the rear and found a door marked 'service entrance'. It was unlocked. She glanced down at the black box. The light was flickering faster. She dropped it into her purse and entered.

In contrast to the front rooms of the restaurant this end of the building was well lit and alive with kitchen people and the sounds of food preparation. She paused just inside the door until a small black man under a very tall white hat cocked his head at her and asked her what she wanted.

"I'm looking for a job" she said, thinking how odd it was. The hat was fully half as long as he.

He stretched to his full height which brought the bottom of his hat just above her shoulder. He looked her up and down. She wondered how the hat stayed put. "Serving or hostess"

"Do I look like a serving girl?" She admonished.

His wide smile was appreciative and showed many white teeth. His voice carried a lilt from the islands. "You could be many tings Madam, but, no, a serving girl you do not look." He turned, pointed through the swinging door, "Take a left, go up the staircase and to the right you find Mr. Vigus, the Manager in his office. I would knock before I go in."

She made her way through the kitchen help to the swinging door. Each paused, watching her walk and all but one suppressed a whistle. She noted there were no women in the kitchen. She went through the swinging door and paused at the foot of the stairs long enough to raise Jason's box out of her purse. As she took the first few steps the light stopped flickering. She took the stairs eagerly.

His door was plainly marked. "H. Vigus, Manager."

She knocked.

"Come"

She opened the door into a spacious office set off with leather chairs, some hanging flowers and a spectacular view of the inner harbor. Muzak plied the air with the muted strains of Montovani. A middle aged, dark haired man with hairy arms sprouting out of his flowered sport shirt looked up.

"Mr. Vigus?"

He smiled. "Yes. What can I do for you."

She favored him with an engaging smile. "I'm Christa Bonamay. The grapevine tells me you may be looking for a hostess."

He rose and walked around the desk with his hand outstretched, "Well, I am always looking for good employees." he said expansively with heavy emphasis on the 'I'. At that moment the phone rang and he answered it. He motioned her to a chair by the window and put the phone down after a very short conversation. "Will you excuse me for a moment, Miss Bonamay, I have a small crisis in the kitchen -- I'll be right back ."

Without waiting for her answer he disappeared through the door into the hall.

She immediately plucked Jason's box from her purse. The light was implacably red and did not waver. She had found the source.

A subtle change in the air made the short hairs on the back of her neck tingle. She was looking at her arm and saw the tawny down there raise and stand on end. At the same time she was vaguely aware that Muzak had changed its selection and the room was filled with a mild hum. She caught movement out of the corner of her eye and glanced up to see the awesome apparition that was Luci begin to materialize. A chuckle, deep as a double bass drum, came from his throat. He swung toward her, slowly as though turning in an underwater ballet, his massive right arm rotating up from his shoulder.

Without thinking she slid out of the chair getting herself below that deadly finger. She hit the floor scrabbling as a lightening bolt tore the back out of the chair where her head had been split seconds before. She made the door on all fours and launched herself across the hall as the top of the door disappeared amid a thunderclap of smoke and fire. She scuttled down half the staircase on all fours then righted herself near the bottom. The giant laugh followed her but, to her relief, she did not see Luci come through the smoke.

80

The front door was locked, and top and bottom bolts were set. Terror lent wings to her feet and she hit the swinging door into the kitchen in full flight. It slammed open knocking a very surprised Mr. Vigus backward and spinning over a food table. She raced through the kitchen with the vision of his bloody face, mashed by the force of the door and made a mental note to never eavesdrop behind a swinging door. The kitchen personnel made no effort to stop her. She burst into the sunshine and stopped running only when she stood in the median waiting for the opposing traffic on Light Street to clear. She did not remember crossing the first three lanes.

She got Cray on the C-fone and Jason picked up. His first question in the middle of her explanation was, "Are you alright?"

"Yes -- at least I think so. Luci missed me and no one is following me now."

"Fine, hop in the chopper and come on home, we need you here."

* * * * *

CHAPTER 16
Monday 09:00 EST; 14:00 GMT; 16:00 Athens, Greece. - 8 hours

The Airbus turned into a long glide and dropped gently through the smog to the landing strip at Hellinikon Airport in Athens. From a city that only held 5000 people in the middle 1800s Peter now looked out upon a teeming metropolis of four million people. Peter debarked before the First Officer could make it to the cabin so that he would not have the entanglement. He walked through the concourse carrying his small bag and entered a coffee shop which afforded him a view of the cab stands and the arrival areas. Wanting to simply mark time, he ordered coffee and unobtrusively watched the Cabstand. As in all airports the cabs were drawn up in a long line each getting his turn at the front of the line. After fifteen minutes of casual observation he noticed one cab at the rear of the line that never seemed to get any closer to the front. It was sitting slightly to the side of the cab line at the rear and stayed there. The windows were tinted just enough to foil his sight but he could tell the motor was running. Each time a customer was picked up the line forged ahead one cab but this cab never moved. He noted the markings on the cab. After at least five cabs had picked up their fares he paid his bill, took his change in Greek money then stepped out to the head of the line where a porter waved him into a cab.

The cabbie turned in his seat, "You American?"

"Yes."

"I spik American berry well -- know the best places -- the best girls..."

Peter waved him off. "Okhee[18], just take me to Hadrian's Arch." Peter bent over as though to tie his shoe and pulled his locator from his pocket. It was pulsing gently but not insistently.

With true Mediterranean fortitude the cabbie shrugged his shoulders and wound his way out into the noisy traffic on

---

18No!

Vouliagmenis Boulevard. As he made a right into the flow of traffic Peter turned his head just enough to see the mystery cab round the end of the line and also enter the flow of traffic. Interesting, he thought. Evidently someone knew I was coming.

Immediately to the sides of the busy thoroughfare Peter saw the trademark multitudes of small billboards with perhaps twenty identical ones at one time in one spot. Ground level, but perhaps ten feet high, the effect was a repeating montage of color, shape and feature that stunned his senses as they spread along the road, some for a hundred yards or more.

Legend has it that when the Gods dug out the Mediterranean they piled the rocks up and later called that Greece, and Peter believed it. The sparse, listless land raised away from the sea under the blistering, bright late afternoon sun and sprouted rocks in all the drab colors of the universe. Compounded by builders who were restricted to concrete, brick, mortar and stone as their only building materials, one was instantly struck by the sameness from block to block. There were practically no trees except Eucalyptus whose leaves collect dust and turn the color of the landscape so there was nothing to absorb the searing sunlight. Peter hastily donned his sunglasses. The only wood he could see was the scrawny ribs used to form the cement. Dry rotted, worm eaten, looking like all the world's woodpeckers had attacked it and then thrown it up on the beach as driftwood he could see they used it only for forming difficult areas like steps, stanchions and pilings. He suddenly realized he was in a country whose forests were history a thousand years before the Pilgrims landed on the rock.

He turned as though to witness the hillside and surveyed the traffic around and behind. The mystery cab was not in sight, but he felt its presence. His gaze was drawn left up the rocky slopes above the highway. One street above the busy thoroughfare residential took over and houses marched up the hill. They became larger, more prestigious as they perched higher above the traffic. He saw one with what was unmistakably a wooden door -- a very ornate, decorated door. . When questioned, the voluble cabbie explained that only the very rich could afford to import the wood and that was probably teak which would weather well and if the man was very, very rich and you went inside there would be much custom wood and delicate filigree.

The closer they got to Athens the worse the traffic became, and everyone seemed to have a horn. At times it was cacophony but the cabbie was oblivious, wending his way in and around the clusters of small, darting cars. Most of the traffic was made of compact cars with an occasional limo or large Mercedes type except for the cabs which were mostly older American Fords and Chevies.

As they neared the town the cabbie pointed forward, up and left and Peter gasped. All his life he had heard of the beauty, symmetry and simplicity of the most perfect building in the world, The Parthenon but he had totally forgotten it in his hurried trip. The awesome outcropping, The Acropolis, on which the magnificent building sat jutted abruptly into the sky from the center of the Plaka[19] like a huge, ugly wart. But the encircling three foot thick walls flowing upward from the shank of the rock were a thing of massive beauty besides providing a protective, impenetrable sheath around the top acreage. The two thousand year old Parthenon protruded above the walls, dominating the skyline -- reclining, brooding. Still at least half a mile from the monument he could see it was the most prominent landmark in this ancient city. He could see the tops of the massive columns holding up the vaulted roof and got a full side view as they waited in the traffic. It was just like the postcards sitting there against the bluest of skies.

They came to a dead stop in the traffic. Ahead and to the right he could see the corner of Hadrian's Arch. He tapped the cabbie on the shoulder, held out a number of drachma bills from which the driver delicately extracted one. Peter grinned then gave him another one of the same denomination which produced a larger grin in return.

He stepped quickly over to the sidewalk walked around the corner, stopped and looked back. There was no apparent pursuit.

He glanced at the locator. Its pulse had quickened slightly. He rotated it in the palm of his hand and watched the pulse quicken as he turned it dead on the Parthenon. The distance was still near five hundred meters. He walked toward the Arch and Lysikratous Street beyond.

He rounded the corner and stopped to look through the chain link fence to his right. It was the remains of the Temple of Olympian Zeus.

---

[19] Old Town

Starkly white, glistening eight foot diameter granite columns still standing eighty five feet tall in the afternoon sunlight. One had fallen some time back in history and lay stretched toward him, its massive many fluted, five ton rounds gleaming in the sun, pointing at him like a giant's skeleton finger.

He turned to look at the double level of the arch that Hadrian had had built in order to herald his victorious entry into the city. Of all the many works the Roman Emperor had constructed in the second century in Athens this was the only survivor. Standing in a certain spot you can frame the Parthenon in the upper background and he recognized the famous picture immediately. Jostled he stole a quick look at the milling tourists but detected no danger. He skirted the traffic across Leof Amalias and entered the thin, upward winding Lysikratous which terminated near the base of the Acropolis.

A marvelous smell assailed him and he realized he was ravenous. With the locator's pulse strengthening he wound his way upward through the crowd just following his nose. The street was perhaps thirty feet wide and flanked by sidewalks which were engorged with small cars parked across the walk like suckling piglets with their noses toward the buildings. This left in the center a narrow lane through which the crowd flowed in all directions, occasionally parting for yet another small car to weave its way through one way or the other. The concrete buildings rose smoothly two stories on both sides of the street. Their walls were broken occasionally with iron balconies, most covered with wandering roses. Small windows pocked the second story walls and the bottom floors were punctured by small doors that led into shops. A glance at the forbidding walls brought a tingle to the back of his neck. What a great place for a sniper, he thought. As he glanced toward the Acropolis he could see the street rose slightly at the upper end as though paying homage to the great rock protruding above it. Having no choice but to move upward with the flow, he ignored the tension and began to pick up the holiday feeling from the crowd. He walked past the shops which were selling everything imaginable including gold, leather and woven articles, clothing, glass, jewelry and the ubiquitous, nude, unmistakably male carvings sporting full erections. He grinned and thought he would like to have one that size in the same relation to his body size. He moved on still following his nose.

Half way up the street he homed in on a Dutch door. Behind the door was a hot, dark little room, no more than a deep closet in which a little old gnarled crone wielded a spatula into some frying oil spitting over a hot plate. She wore high topped shoes, a tattered, work worn print dress covered by a grease spattered apron and had her hair tied back under a babushka. The aroma was delicious. He looked over the head of a customer waiting impatiently and watched her concoct her little miracle. In her left hand she had some sort of flat, tortilla like pita bread into which she laid a wooden shish kabob holding cubes of meat and various kinds of veggies which had been deep fried in the sizzling pan. She clamped her left hand on the wafer of bread, withdrew the skewer like she was pulling a small sword and deftly wrapped the thin bread around the veggies and meat to finish up what looked like a skinny burrito. The smell of spicy meat and seasoned, seared veggies was delightful and overwhelming. The sign said 'Souflaki, 100 drac'[20]. The customer in front of him handed the crone some money and departed happily.

Peter held up two fingers and pointed to her handiwork.

* * * * *

20Approximately 30 cents American

CHAPTER 17
12:00 EST; 17:00 GMT; 19:00 ATHENS   - 5 HOURS

Christa keyed in for L'Aerie, got Cray and was immediately picked up by Jason.

"Cray shows you over Bowdon ,NC now" He said. "Can you confirm?."

Christa pushed the bird's left side down and surveyed the ground. "That's probably it down there. I'm about 135 miles out."

"Good you're less than an hour -- we need you."

"Is everything progressing?"

"Yeah, the Gov is getting ready for the big meeting. The death toll is over ten thousand and still mounting although is seems to be a little slower. Peter has landed in Athens..." He continued telling her of Luci's attack on Peter in the plane and assured her he was alright.

She calmed down a bit. "Do you have any further information on how they are doing this?"

Jason's voice was far away, contemplative. She could tell that, even as he spoke, part of his mind was rummaging around, looking for an explanation. "It has to do with sound but we don't know how yet. He always comes in with sound. Come on in, Paul and I will be waiting for you."

Christa closed the circuit and checked the ground below. The pulpwood smokestacks of Bowdon were fading back to her left and she traveled over sparsely populated, thick pine forests splintered by isolated roads. Boredom set in and she punched the radio looking for news. Country music flooded her earphones at first. As she enjoyed the twang suddenly she realized the beat had changed to a steady hum that grew louder as she recognized it. There was movement in the co-pilot's seat. She turned and her eyes went wide with horror. Lucifer materialized beside her and there was a blinding flash....

* * * * *

The old lady looked up at him and went about the business of making two Souflaki which she placed in front of him. "Un Beerah?" she asked.

He nodded and she pulled a cold beer from a cooler. He extended his hand with money and she daintily extracted 500 drachma and began counting his change. He bade her 'Okhee', stored her broad curtsy and wider smile for a sadder time and moved off through the crowd, munching on a Souflaki. It was a taste experience. Succulent cubes of lean, spicy lamb alternated with tender chunks of vine-ripened tomatoes, crisp celery, and sweet Mediterranean onion, all deep fried in olive oil and topped with a sprinkle of mixed spices. He would have had two more but was half the block above her doorway when he polished off the last of the second one and dropped the empty beer bottle in a convenient trash can.

He glanced quickly at his Locator. It was now beginning to vibrate in his hand and the light showed almost a steady pulse. It showed twenty five meters. He walked quickly up the slope and stopped directly in front of Poulos' Oozehree[21]. The Locator went crazy. He turned it off.. He stepped through the open door and paused until his eyes could adjust to the dim light. It was like stepping into a candle lit cave. The eyes of a dozen locals appraised him as he stood in the inside shadows. He scoped eight customers at the bar and the surrounding tables plus a bartender and three serving girls in a clutch at the far end of the bar. The way the girls were dressed he was sure they served other functions as well. The tables were lit with softly flickering candles casting eerie shadows on the darker walls. Were it not for his purpose this could be mistaken for your friendly neighborhood watering hole. He took a seat at a table that afforded him a view of the front and back doors simultaneously.

One of the girls immediately detached herself from the clutch and flexed her way to him. She smiled -- a toothy smile, missing only one cuspid on one side. She wore a printed sheath dress which left her shoulders and arms bare and exposed long, slender tanned legs below mid thigh. As she stood in front of him his Locator inside his pocket

---

21 Poulos' Bar

began a slow vibration against his leg. By God, he thought, she's wired. He wondered where it was.

She spoke in heavily accented English. "Bon soir, you are Americain, no?"

He grinned up at her although she was not much taller than he while he was sitting. "Yes, is it that evident?"

"A compliment senor. Your clothes definitely are states -- your haircut too. What would you like une beerah – something to eat – a girl perhaps?

"Let me try this." He pursed his mouth and spoke very slowly and distinctly. "Thah eethehlah meeah beerah, pahrahkahlo -- Ahlfah?" [22]

Her smile got broader. "Meelahteh ehleeneekah"[23] she shot back.

He looked at her helplessly, smiled and said slowly, "okhee" [24]

She shrugged amiably and turned to leave his table. He asked, "Poo eenah ee tooahlehtehss?"

She regarded him quickly for a moment, suspicion rising in her eyes, then struck a finger toward a dim hallway off the end of the bar.

He followed her across the floor, breaking off to enter the hallway where he found the traditional signs indicating men's and women's restrooms. He noticed the end of the hallway had an 'Exit' sign. He swung cautiously through the proper door into a somewhat brighter room to see first a sink at his immediate right topped by a large dim mirror then three urinals behind it on the same side, faced on the other side by four stalls with partial, swinging doors.

His Locator told him he was not alone in this room but no feet were in evidence below the stall doors.

He moved warily to the rear and took the position in front of the corner urinal but did not unzip his fly..

He cocked his head slightly. A wee squeak of a rusty hinge betrayed the rush telling Peter to stoop and turn.

His assailant was clearing the door from the nearest stall, a wicked eight inch knife pointed from his right hand toward Peter. The

---

22 "I'd like one beer, please. An Ahlfah."

23 "Do you speak Greek"

24 "No" (Reader please understand that these are all phonetic renditions of the original Greek).

knife blade was like an ugly snake, darting, parrying, then darting again. Peter extended his hands palms out and the knife wielder stopped six feet away -- on guard.

"What do you want?" Peter asked casually.

The knifer was in his mid twenties, dark and muscular. Peter's confident reaction to his knife attack had temporarily confused him. He rapped out a command. "I am the one with the knife senor, I will ask the questions."

"Ok, what do you want to know?"

Suddenly the doors of the three remaining stalls opened and Peter's confronter was joined by three clones jacking rounds into their automatics. All four were dressed in rough peasant clothes. The tallest of the four stepped forward waving his gun and said "I will ask the questions."

Peter cocked his head toward his interrogator. The man was almost as tall as he, heavier built, swarthy from too much Mediterranean sun, his lips curling behind a heavy, black mustache. He had a spotted red kerchief knotted around his neck.

His English was stilted but recognizable. "What do you want in this place?"

"I'm a tourist. I stopped in for a beer."

"Liar!" he spat waving his gun at Peter. "You are no more a tourist than I, Mr. Chaney, you came here trying to trace Lucifer. Admit it!"

Peter kept perspective on the knife wielder who was edging slowly forward. He was almost in striking distance.

The leader spoke again. "I ask you again, what are you doing in this place?"

The knifer shuffled half a step closer. Peter judged the distance and bought some time. "Is this the way you treat American tourists who come into Athens?

Thinking Peter distracted with the conversation the knifer took the ill fated opportunity to again decrease the distance between them. He failed to note Peter's long arms and to realize that the critical distance for Peter was six inches longer than his.

Peter's left hand shot out and took a steel rimmed grip on the outside of the knifer's knife hand. Taken totally by surprise by this tall, soft spoken stranger, the knifer stiffened -- but too late. Peter jerked him forward, off balance, snagged the knife hand with his right and drove the knife back up into the thug's belly.

Defenseless prey is not supposed to turn on its armed attackers, particularly with such determined ferocity and it took the leader too long to sort it out. He missed his wide open shot and when he did fire it was into the dying body of the knifer which Peter was propelling into the three others with all the gentleness of a charging rhino. The body sent them sprawling, shots flying into the ceiling and walls. A haze filled the air and the roar of the shots was deafening in the small room. He stayed low, hacking, slashing and punching at what ever part of their anatomy he encountered. In just moments there was no sound but muted moans. Peter crouched in the corner below the haze. A quick inventory gave him bumps and bruises but nothing serious. His ears were ringing but not so much that he didn't hear the sound of feet rushing down the hallway. He stepped up against the wall beside the door just as the door opened. The bartender stepped through with a baseball bat and Peter clipped him sharply in front of the ear and heard the jaw snap. The barman dropped instantly without a sound. No one else was moving on the floor nor could Peter hear any moans.

He slid through the door, turned right and edged through the 'exit' door into the velvet arms of the early night.

* * * * *

Cray interrupted Jason. "Mr. Chaney is on the scrambler."

Jason grabbed the phone. "What's the latest, Peter.?

He listened for some moments as Peter recounted the episode in the bar. At last he could contain himself no longer. "Are you alright, Peter?"

"A bit out of breath, but, yes, I'm alright."

"Where are you now, " Jason asked

"Somewhere in the rocks just below The Acropolis."

"That's opportune because we are getting sporadic transmissions and Cray and the G.P.S. place it as coming from the Acropolis and possibly from the Parthenon itself."

Peter looked down and along the slope below him. In the moonlight he could see massive chunks of stark white sculpture of all kinds. Granite Rounds, granite chairs, obelisks, pieces of column cast around like junk around a giants worktable. "There's a monument storage yard below me but it looks like I'm about two hundred yards from the steps leading up to the entrance. Won't take me long to get there..."

"Good. Keep in touch..."

"Wait! Jason. How's Christa?"

It was Peter's turn to stand transfixed as Jason told him of her encounter in the restaurant. Peter cut him off. "You know that she's alright? You're sure?"

"Yes, Peter. She just called me a few minutes ago. She was a little under an hour's flight out and on her way. She'll be here shortly."

The relief was evident in Peter's voice. "Good, how's Paul?"

"Alive and well."

Peter grunted. "I'm on my way up to the Parthenon."

Jason's voice came from a long, long way over the satellite and sounded every bit of it. "Use extreme caution, Peter. We've already found that these people play for keeps -- I don't want anything to happen to you."

* * * * *

CHAPTER 18
12:30 EST; 17:30 GMT; 19:30 ATHENS; - 4.5 HOURS

Cray picked up Christa's chopper on maximum at about ten miles out. When queried by Paul there was a pause then she came on line and requested permission to land. He could hear soft music playing in the background. He thought it odd that she was so formal but gave her official permission. He zeroed in on the cockpit and kept the resolution on fine even as the bird circled to land but couldn't tell much behind her helmet. The landing was perfect.

"She's awfully quiet, Jason. That experience at the restaurant with Luci must have really shook her up."

Jason looked up from his monitor. He was getting the latest body count -- it had now jumped to over twenty thousand. "She'll be stepping off the elevator in a moment -- maybe you can cheer her up."

Cray suddenly chimed in. "There is an odd aberration in the electrical aura in the building."

Jason was about to query Cray in more depth but the elevator door at the far end of the CPU opened and Christa stepped out into the room. She smiled at them and she positively glowed. In fact when Jason looked close to her but not at her he could sense -- more than see -- an aura -- a shining about her.

"Christa," he said making no move to hug her. "Are we glad you're home. Talk to me about Baltimore."

Whoever had manufactured her almost pulled it off. Her speech patterns were identical, the vocabulary was there but the rhythm gave her away. And the essence was different. When he looked directly at Christa it was obvious that she was not there in the flesh. There was an other worldliness, a falseness, an almost plastic look to her. Her voice inflections were perfect, her moves were pure Christa but the apparition in front of him was simply that -- an apparition. A very skillful, awesome technology had produced a hologram that had almost passed for real.

Paul skirted around her, standing on one side, then on the other, studying her, listening to her, meanwhile very careful not to touch her. She reacted to only one man and one conversation at a time and ignored whoever was not in front of her.

"Christa!" Jason cut in sharply. "Tell Cray to get you the latest body count for me."

She smiled and said "Cray, get me the latest body count for Jason."

Cray did not respond.

"Tell her again."

Christa said it the second time and Cray ignored the command.

Jason said, "I thought so, you are not Christa -- what have you done with her -- where is she."

Paul raised his hand to strike the apparition and Jason yelled, "No Paul! Don't touch her. She may electrocute you..."

Suddenly he stopped and both men stood spellbound as Christa began to morph -- into Luci. The monster devil's head showed first, grinning evilly, chuckling deep in its barrel chest as the figure flowed downward fading out Christa and grew to its massive proportions.

Paul dove behind a desk. Jason took several quick steps backward and stopped.

"You are not frightened, Mr. "A", Luci rumbled.

Jason drew a deep breath. "Yes -- but I'm also intensely curious. Who makes you? Where do you come from..." Then he became aware of the beat. The music was there and revelation hit him. He jerked his earphones off, expecting to watch Luci disappear -- but the monster was still there.

"Did you think that would make me go away?" Luci rumbled and the glassware in the CPU rattled. He made a sound like a hurricane eating up your backyard, or Apollo Thirteen taking off through the bathroom window -- a sound so big that it bypassed your ears and simply flowed through your head -- so low it made your teeth ache.

Jason knew he was looking at a simple act of physics -- just an electrical phenomenon but the sheer enormity of the accomplishment was staggering.

Luci rose to his full height, almost touching the vaulted ceiling of the CPU. Jason knew what was coming – always the bolts came when the monster was standing upright.

"Cray," he screamed. "Shut your modems down!"

There was no discernible change in the sound in the CPU but Luci suddenly stopped moving and quickly began to fade -- and was gone in seconds.

Jason danced a jig. "By God, Paul, he has to have music. Cut off his music and you've cut him off at the pass."

A sheepish Ashton crawled from under the desk and wiped imaginary dust off his pants. He looked around. "Jason, Where's Christa?"

Jason stopped dancing, and said "Cray, turn on all your modems. Where's Christa."

"Christa is in the chopper"

"Cray. Is she -- alive?"

"All her vital signs are normal. She appears to be in deep sleep."

"Go get her Paul and let's talk to her."

* * * * *

The bird was crouched on its customary berth just scantily over half a rotor from the hillside wall. Whoever had piloted it to this landing had known exactly what they were doing.

Christa was curled in a heap on the floor of the chopper with her back to him. The air around her smelled vaguely of something burnt but there was no smoke. She was apparently not harmed, at least on the outside. The bird had landed safely so Paul wasted no time inspecting it. He leaned over her back, put a hesitant hand on her shoulder and shook her gently. She did not move.

He leaned closer to her face and shook her gently again. "Christa -- Christa! Can you hear me?" This time she moaned and her eyelids fluttered. This was the closest he had been to her and the clean, light fragrance in her hair made him want to bury his head in it. She turned slowly like a big cat and in a moment he was gazing into those bottomless blue eyes which opened fuzzy and vague then slowly focused on his. He was so happy he almost kissed her on the tip of her nose but suppressed the urge. "Are you alright?"

She began sitting up. "I -- I think so. What happened! The last thing I remember..." She spun on the floor and looked wildly at the passenger seat... "That god awful monster was sitting there grinning at me..."

Paul patted her shoulder. "Well, he brought you and the chopper back here and tried to kill us but Jason stopped him..."

With his assist she got to her feet, shaken but undamaged. "Help me get downstairs. If Jason has made Luci I want to hear about it."

* * * * *

"Well, I am certainly glad to see you back in the nest in one piece and no burns," Said Jason.

Paul and Christa had just stepped off the elevator into the CPU. Paul had been more than a willing supporter for her on the ride down but now she stepped away from him and regarded Jason.

"Do I understand Paul to say that you stopped Luci?" She asked without preamble.

"Temporarily thwarted is a better term" Jason grumped and described the terms of his encounter with Luci. He also described the Christa hologram. Christa listened with bated breath.

When he was done he said "Now tell me? You are probably the only living person that has seen the apparition twice and still lives. What is your impression?"

Christa paused for a moment, digesting what Jason had just told her. She slumped down into a chair. When she spoke it was as from deep thought. "If the original encounters displayed the limits of Luci then they have upgraded the technology. To create an abstract figure like Luci requires a template of some kind and once you've done it you simply redo it as often as you like. But in my case they sent Luci into a live, real time situation and created something totally foreign, me, literally from scratch and on the spot."

"Yes, Now we're getting the measure of the people we're dealing with and I think we may have an adversary worthy of our mettle."

"Any idea who these people are?" Paul asked.

"No, but I think I know what their next step may be."

His words electrified both of them. They turned to him and waited.

"They know who we are, where we are and how to get into our midst. The next step is to send us a virus to disable Cray."

Christa scoffed. "We have the finest anti virus protection known to man."

Jason grinned. "Under ordinary circumstances I would agree with you but we are not dealing with a known quantity here, and judging by what we've seen they may have other aces up their sleeve."

"What do you suggest?" Christa asked.

"I want a dump file for all incoming. I want you to program it so everything that comes in dumps there and stays there. I want that dump file on a separate, small CPU unconnected to Cray. Then we can have Cray inspect the incoming and when we're ready to import we can do it manually and this includes phone calls."

"How about Washington?," Paul asked.

"We have fifteen incoming telephone lines and three direct satellite dishes. We'll block fourteen of those lines plus the three dishes and let Cray inspect everything that comes in through there. The number for the fifteenth line will be given only to four people. The President, Jed Foley, Special Agent Malawi and Peter. Those four can raise us direct without getting mixed in with the rest of the data..."

"How soon do you want this done?" Christa cut in.

"Now. How long will it take you to write the program?"

"Fifteen minutes of solitude -- the hardware is in the storeroom."

"Paul and I will assemble the hardware," he said getting up, "Shut Cray down and get busy."

"How will we get anything out?" Paul asked as they made their way to the storeroom.

"Cray will assemble packets of information. When one is ready Christa will open the lines for a nanosecond or two, just long enough for a burst and when we've got the message out she'll shut the lines again."

"How long will it take Cray to make her inspection?"

Jason was thoughtful. "I'd guess a delay of fifteen seconds probably. Too much to carry on a real time conversation but sufficient to handle any other needs."

* * * * *

The quiet speaker summoned the resting pilot. "Armand, come into the control room." Dark haired, slender, he moved like a stalking cat down the dim hallway and into the brightly lit control room. Several technicians sat at consoles in front of the large tinted windows that viewed a huge cavern perhaps fifty feet deep and over a hundred feet long. The walls and ceiling of the cavern were rough cut from the black rock and crisscrossed with hundreds of electric lines, water pipes and conduit. The cavern was so adroitly and brightly lit there were no shadows. In a line in the middle of the room, approximately fifteen feet apart, sat three virtual reality stations looking like giant hollow lattice- work balls. Behind each was a massive wall screen which depicted the situation in which each Luci was engaged. Each ten foot ball was occupied inside by a figure suspended by hissing jets of air. They rotated freely within the air cushion like training astronauts within a gyroscope. Each figure wore a dark snug pressure suit similar to a wet suit. The suits were pockmarked by a myriad of sensors attached to fine wires running from various parts of the suits to the outside frame so that each minute portion of the suit could send a signal. He smiled as he saw one of the figures raise his right arm in Luci's characteristic booming salute and the lights dimmed slightly in the control room.

He approached the one man who was on his feet and clearly in charge.

"Commander", he said with great respect.

The tall man bent over a console was middle aged, fair skinned with a dark mustache. When he straightened and turned he was wide enough to be intimidating and tall enough to look down on the young pilot. He was angry.

"Armand, you are my best pilot but Amador forced you out -- how?"

Armand shifted uneasily. "That old man is extremely quick and he has figured out that it is the beat that gives us entry. I was not prepared for that." he ended lamely.

The Commander's eyes snapped. "Why did you not kill the girl?"

Armand was still defensive but his manner quickened. "Sir, that was my first experience in astral projection and I needed her vibes to maintain the form -- but I couldn't make her look solid enough to pass Amador's scrutiny."

The Commander pursed his lips. "A minor problem which the programmers are working on as we speak." he paused then his manner became stern. "The next time you have occasion to do that you kill the subject first -- we want no survivors. Is that understood?"

"Yes Sir."

The Commander turned away then had second thoughts. "Also you must speed up the bolt motion. If you had been faster Chaney would be dead now instead of giving us fits in Greece. You are dismissed."

"Yes Sir."

* * * * *

CHAPTER 19
1 P.M. EST; 18:00 GMT; 20:00 ATHENS; - 4 HRS

Peter slid quietly down the alley behind the bar, checked carefully around the corners for anything unusual and, sensing nothing, eased himself into the light crowd walking up toward the Acropolis. Prudence told him that he should heighten his pace and he began to thread his way through the crowd ahead. The crowd milled around a bit as they passed the impressive Monument of Lysikratous, then turned onto Epimenidou Street. As he passed the ancient Karaghioza Theater with its' famous cartoon wall, he took a steep set of steps which brought him out onto Thrasillou Street and thence down to Dionysiou Areopaghitou Avenue. He drew back into the shadows at the base of a street light and gazed long and hard at the crowd above him coming down Thrasillou toward him. He could not detect anyone other than innocent tourists.

He turned and started up the gentle slope of Areopaghitou. On his right an opening appeared in the fence behind which he could see another of the giant's play yards. Hundreds of pieces of granite and marble, some as large as a small car pitted the hillside leading up to the Stoa of Eumenes and the Theater of Dionysos. Another time he would have gladly stepped in and embraced the mystic of two thousand year old pieces of an unknown puzzle but tonight was not the night. He continued up the block past the grand marble stairway entrance to the Herod Atticus Theater to the entrance to the Acropolis and began to climb up the marble steps leading up to the Acropolis. In spite of his excitement he had to caution himself to take them one at a time so he would not draw unnecessary attention.

Dusty green trees, olives and carob, embraced the broad glistening steps and almost formed a tunnel up the side of the Acropolis. Near the top he handed over money, received his ticket and change and followed the crowd up through the massive stone portico guarded by two oversize granite lions, one reclining on each side of the steps that led up to the temple of Nike Apteros.

It was an ideal layout for a stalker and it made him pull back to the side of the crowd to study the probabilities. To his right was the massive, impenetrable stone wall of the temple which rose perhaps forty feet straight into the warm early evening air. To his left roped and marked 'out of bounds' because of restoration and in front of him were the famous, very steep but very beautiful, two thousand year old marble steps leading up through the Propylaea and into the Acropolis grounds. The steps rose two or three stories above him in terraced groups and were teeming with people -- but no one seemed to be waiting for him so he took the steps apprehensively, two at a time.

The roof of the Propylaea had long ago been lost but the massive, round columns on which it rested still stood. Eight to ten feet in diameter they rose majestically on his right and left and reached into the velvet sky. He passed them and stepped into the main yard atop the Acropolis. Again caution drew him into the shadows to judge his surroundings.

A stony path led up towards the Parthenon which stood about one hundred yards up and slightly to his right. Opposite it, to his left he could see the Porch of The Caryatids, the famous female statues which are the only known pillars made in the human form in Greece. At the far end, and well above him, he could just see the stone battlements from which waved the blue and white national flag of Greece.

In between was another field massed with a conglomeration of granite and marble pieces of every size, description and hue. Some inscribed, some not but as he passed them he noticed that all were numbered like giant pieces in a jigsaw puzzle.

As he drew near the Parthenon he saw that it was under massive repair. Scaffolding stood like gaunt ribs around all four sides of the roofless structure. A crane stood in the middle of the floor and extended high above the open roof line. All was enclosed by a heavy rope which dangled signs in three languages saying 'no trespassing'. As he watched, several workman, some in shirts and shorts, others in just shorts, all wearing heavy boots and carrying lunch buckets wound their way through the huge blocks on the raised floor of the Parthenon, descended a ladder and walked wearily past him.

The imposing Parthenon stood brooding atop a base perhaps twelve feet high. Built by master craftsmen, first to dominate the high field of the Acropolis, then the whole of the city of Athens, no intention had ever been more totally fulfilled. It was a structure that spoke to man's lofty ideals, to justice and to equality. Momentarily forgetting his purpose he stood transfixed before steps glistening in the waning rays of the sun, worn thin, covered with a polished patina reached only after being caressed by countless thousands of feet over the past two millennia.

His locator beat a tattoo against his thigh. He pulled it out and said, "Yes, Mr. "A".

"Peter, I know where you are standing, and I know what you are thinking but there is a transmission originating less than fifty meters from you and it needs your immediate attention. Cray places it smack in the middle of the Parthenon."

Peter walked around to the side of the Parthenon. He stopped between it and the Acropolis wall. No one paid attention to him. "The middle of the floor is jumbled with huge pieces of granite and marble. There's a crane up there but all the workman have supposedly gone home. I'm blocked out by a rope but I can go over it."

"Good luck, keep us posted."

He pocketed the locator and immediately stepped over the rope as though he had every right to. One or two tourists glanced in his direction but assumed, since he had been using a telephone, and looking up at the structure, that he was somehow part of the operation, and did not question him. He stepped carefully up a convenient ladder and disappeared into the maze of blocks, most of which were taller than he.

After he passed the second corner in the labyrinth all sound went dead. The sounds of the world disappeared -- seemed to flow into the rock and just -- disappear. He could see the tepid sky overhead, now speckled with new stars because the rocks blocked all artificial light from his position, but he could not hear a sound. He walked very, very softly and felt the air grow heavy with the tension of the hunt.

The locator vibrated against his thigh He put it to his ear.

Jason said quietly. "You are less than seven lineal meters south of the transmission. Can you hear them?"

Peter quietly tapped the Mike. once with his index finger.

"That means they are transmitting non voice. Try to get the camera in range so Cray can record. Stay close to them, Peter and see if you can drop a tracking pellet[25] in one's pocket and we'll find out where they go -- but be careful."

Peter tapped the Mike. twice and manipulated buttons and switches on the bottom of the locator. He slid quietly up to the corner of the block he was standing by, bent over, held his breath and eased the camera around the corner at waist height. It took a moment to position it but then he saw that inside the wall of larger blocks there was a semi-cleared space in which several smaller blocks were scattered. Twenty feet from him three rough cut men, dressed like the workmen who had left earlier, nervously huddled over the table top of one rock. On it was a laptop computer, open and operating. One man was punching the keys while the other two kept furtive watch. Peter clicked the silent shutter several times then eased back from the corner, began to breath again and started his noiseless retreat.

* * * * *

"Would you look at that," Paul exclaimed. "Clear as a bell."

"I must say I'm rather pleased," Jason said modestly, "this is the first time that camera addition has been tested in the field and the pics are very sharp."

He manipulated the dials and zoomed in on the laptop. It was a cross view of it so he was not able to see the screen but could clearly make out which keys the man's stubby fingers were punching. There were five pictures in all which gave him detailed photos of all three men. "At least we now have some faces to put to the messages they are sending."

"Why isn't Cray reading the messages back to us?"

"They are uploading and storing the packets on the satellite. Obviously they are getting ready for a massive transmission but they will burst it in a few nanoseconds only when they get ready."

"Can we stop it -- or intercept it?"

Jason stopped tapping the keys and looked at Paul. "No and no. Cray is trying to break the password on the packets now but with even her prodigious ability it looks like its going to take longer than we

---

25An electronic device about the size of a nickel that emits a silent tracking signal and can be followed by satellite.

have. The best we can hope to do is to track it from the satellite when it breaks free."

"In the meantime we wait and watch."

"Yes, in the meantime we just wait -- and watch."

* * * * *

Peter dropped off the ladder, stayed in the shadows and finally joined the crowd fifty feet from the Parthenon. He continued to drop back down the hill until he could sit down on a large piece of granite right next to the pathway where he had an unobstructed view up the hill to the great building.

Forty five minutes later he was about to go back up the hill and see what had happened to the trio when he saw them make their way to the side of the Parthenon and drop down to the ground. One held the laptop close to him in a small tote bag.

As they approached Peter he rose and took a few steps into the crowd. As the three men hurried past him he took two or three rapid strides which brought him close abreast the laptop carrier. Without pausing in his stride he dropped a tracking pellet into the thug's side pocket then stopped at the edge of the crowd as though he wanted to take one last look at the scene above him. After a moment he turned slowly and watched the three as they dropped hurriedly down the steps and out the gate into the darkness.

He slowly followed as he keyed in for Jason. "I've just planted the pellet."

"Yes, we're watching the distance between you and them widen now. They are moving very rapidly down the hill."

Jason squinted up at the large screen where Cray was tracking the pellet. "I think -- as we speak, yes -- they are apparently speeding off in an automobile. Why don't you grab a cab at the bottom of the hill, go over to the Palacio and crash. I'll call them and make arrangements and we'll wake you when the trio settles down.

* * * * *

Jason sipped at a cup of coffee and spoke to Christa who had her eyes glued to a large scale map of western Greece. "Where are they now?"

"In the port town of Rafina at a small airfield on the edge of town. I think they'll be flying soon."

"Probably," He turned to Ashton. "Paul, what's the latest on the deaths?"

Ashton was at the third console. "Encouraging news there Jason, The total is up over twenty five thousand but seems to be slowing down."

"Any new stats"

"Yes, there is a definite correlation between literacy and the deaths."

"How so?"

"The higher the literacy rate the higher the death rate."

"Fate is penalizing the literati."

Paul yawned, got up and stretched and started for the door. "Seems so, I think I'll take a break and go get the mail. I guess the Postman is still running in the midst of this."

"The blip is moving."

Jason swiveled his chair toward Christa. "What direction?"

"It circled away from the airport but has now settled on due east."

Jason rubbed his hand over his eyes. It was times like these that he began to feel his age particularly when sleep eluded him as it had last night. Christa making it back safely from Baltimore in spite of her encounter with Luci had given him renewed confidence that they could beat the apparition but the malevolent devil was still very much a deadly adversary and an unknown quantity and Peter was out in the middle of the fray -- with very little protection.

"How far is it from them to the Turkish coast?"

Christa punched a key, "Roughly one hundred kilometers, depending on where you want to land."

"They are in a chopper so they won't stray offline very far. Give me a list of what's in their path."

Christa drew a box over the area and flicked a key. The large map of the small area came to life. Only one island lay directly in the path of the chopper.

"There's a tiny island named Piros just two miles off the Turkish coast. It could be their destination -- Cray,"

"Yes, Miss Christa?"

"Recall everything you have on an island named Piros two miles off the coast of Turkey in the Aegean Sea."

"Yes, Miss Christa."

* * * * *

The fax machine sitting in the midst of the velvet blackness under the single spotlight began to whir. The message was cryptic: PC; Palacios, Athens, Rm. 342. It beeped quietly as the transmission ended and a hand reached from the shadows, picked up the sheet and disappeared back into the shadows.

* * * * *

Paul stepped out of the elevator with his arms full of mail.

"Anything interesting," Jason asked, not really wanting much of an answer but mildly curious.

"Mostly advertising and junk mail. Couple of ads from brokers and three or four magazines." Paul scattered the mail out on the table. "Hey Jason, here's 'Verite', and it's the same issue we've been tracking."

"Don't touch it Paul," Jason snapped sharply,, "let me have it ."

He donned a pair of latex gloves from a drawer. Paul and Christa flanked him and the air in the room turned heavy as he gingerly opened the first page. The magazine was standard size, most pages slick and glossy -- attractive to the avant garde. He turned the pages slowly displaying mostly advertising for the 'under 30's' full of lingerie, high fashion and fragrances. He turned a page to reveal a double truck[26] and one of those folded half pages stuck together by a fragrance. When you rip it open you get a free sample of the fragrance.

Christa reached across Jason's hands and hooked her fingernail under the creased edge. "Want to smell the latest perfume, Paul, they always smell so good."

Time went into extreme slo-mo for Jason. A silent flash of terror tore through him and he caught Christa's wrist before she could sunder the page. He held it steady for a moment, not realizing the

---

26An ad that runs across both pages.

strength he was exerting on her arm, until both she and Paul looked at him and she winced, "Jason, you're hurting me."

He drew a deep breath and moved her hand back from the page, closed the magazine decisively and said, "I'm sorry, Christa, I did not mean to hurt you."

He picked the magazine up and turned around. "Cray!"

"Yes, Mr. A"

"Get me Dr. Ebo at the CDC and put him on the wall."

Christa was rubbing her arm where Jason's fingerprints still showed. Paul was not sure how to take the scene he had just witnessed as he had never seen this side of Jason before. The last thing Jason would do would be to hurt Christa yet he had nearly broken her arm.

The wall screen suddenly came alive and Dr. Daniel Ebo was standing there. "Yes, Jason?"

"Daniel, thank God I've got you so quickly. I think I've solved their delivery system. Do you have a copy of either one of the magazines, 'Verite' or "Young and Lovely'?"

Dr. Ebo spoke in measured cadence, almost a lilt. "We can look Jason, but that is an unusual type of magazine to have around a lab. Do you think there is a connection?"

Jason smiled. "Yes, I do. I think the bacteria is delivered in a 'scratch and smell' ad for women's perfume. I think that when you break the seal you expose the bacteria to the air and when you breath it you later die from it. Under no circumstances do you let any of your people open one of those pages without the highest measures of personal security including wearing your best 'Haz-Mat'[27] suits.

"Noted Jason," Ebo said as he was jotting notes on a small clipboard. "So you really think this is the way it's done?"

"I know it Daniel. It all fits perfectly, just don't let any of your people die from it"

Ebo smiled grimly. "I'll call you back as soon as I have confirmation. If you are right you may have just saved the world."

The screen went dark and Jason turned to Christa and Paul. Christa

---

27Hazardous Materials Suits designed to give the wearer a sealed body space independent of outside influences including its own supply of oxygen.

face was a near blank as he had ever seen it and she was trembling. He patted her on the shoulder as she and Paul both started talking at once. He held up his hand.

"Cray!"

"Yes, Mr. 'A'".

"Get Mr. Malawi on a scrambled line please."

He turned to Christa and handed her the magazine. "See if that is the only page like that -- but don't open it -- then start hacking. I want to know who processes that page and how it's done."

"Mr. 'A', Mr. Malawi is on the line."

Jason sat down and flicked a Mike. on. "Mr. Malawi, Jason Amador here. We've discovered the way the bacteria is transmitted."

There was a grunt from the other end. "Thank God, how?"

"It's in the fragrance pages contained in "Verite" and "Young and Lovely". When you tear it open and bend down to smell the fragrance you get a massive shot of the bacteria.."

He snorted. "Jesus Christ! Everybody smells those things." He paused for a moment then said quietly. "That explains why less literate countries have less deaths. If you can't read you wouldn't be looking at a magazine."

"Yes, it all falls into place when you have the key. The CDC is going to confirm it and you should be getting word shortly from them but I know this is the answer. Will you get word to the President and Mr. Foley?"

"Yes."

"Good, we're trying to locate the company who actually processed that page -- when we have the answer I'll get back to you. Talk to you later."

Jason swung around to face Christa who was literally banging on the keys but Malawi yelled "Wait" and he swung back to the Mike..

Malawi's words tumbled out. "Jason, Chaney's attackers were Tanner's stringers.[28] The CIA made him at the New York Airport and sent word ahead thinking they could get him to talk."

"They don't know Peter very well"

28Part time field operatives only used if desperately needed and then not given positions of responsibility.

"Well," Malawi's tone turned laconic. "For that underestimation Tanner lost three men and the fourth and the bartender are incapacitated in the hospital. If Chaney ever gets tired working for you maybe he'll give us a shot."

Jason chuckled. "He wouldn't fit your profile. Perhaps now Tanner won't keep you in the dark on his operations. Ta."

He turned to Christa. "Any luck?"

She didn't take her eyes off her monitor and kept hitting the keys but said "Keep your eyes on the wall."

Jason shifted his attention to the wall which suddenly lit up with Cray's cryptic comments:

"Both Publisher's used a small print house in Memphis, TN called 'Tru-Print' which specialized in the labor intensive task of brushing the fragrance on the page then gluing it together. The pages for the two magazines were done on different occasions but the delivery receipt and the return bill of lading in both cases were signed by an 'Arulo Metaxis'. Tru-Print is listed as a Delaware Corporation..."

"Is that all,?" Jason demanded.

Christa's fingers did not pause but she tossed her head impatiently. "I'm in the Delaware Secretary of State's Files now -- here it comes."

"President of Tru-Print is Dennis Chillane. Vice President is Thomas Michaels. Off Delaware address for Dennis Chillane is POB 71, Prison Road, Represa, California; off Delaware address for Thomas Michaels is POB 515, Joliet, Il."

She turned to Jason. "You recognize those two addresses don't you?"

"Yes, dammit. Folsom and Joliet. Both of those men were among the first to die of the bacteria on death row."

"Dead end" said Paul.

"No!" retorted Jason. "It's confirmation that we're on the right track. Christa! Put Cray onto both those names to see if there is any correlation besides the fact that they were both in death row. While Cray is doing that I want you to hack Tru-Print's computers and come up with everything they've got."

"Paul, fax the wardens and tell them about the magazine. Tell them to toss their prisons for every copy of the two magazines that they can find and don't let anybody read them when they pick them up."

He turned back to Christa whose fingers were still flying. "A couple of months ago you set Cray to compiling a list of every Bulletin Board on the web, didn't you?"

She nodded without breaking pace.

"Good. Soon as you can, drop a message on each one of them about the magazines with instructions to immediately burn any copies found. Send the same blurb to AP, UPI[29] and Reuters -- the rest of the news services will pick it up from them."

"Tru-Print is not online. I can't hack something that isn't there."

"Cray!"

"Yes, Mr. 'A'."

"Send a message to Mr. Malawi. Suggest you send local police to Tru-Print in Memphis TN and confiscate all available copies of the deadly pages. Caution locals, pages must be burnt without opening."

Jason got up and wandered off toward the kitchen muttering to himself. "Maybe now we can stop this carnage...."

29Associated Press and United Press International

CHAPTER 20
3 P.M. EST; 20:00 GMT; 22:00 ATHENS;   -2 HRS

"My apologies for taking so long Jason -- it took longer than I expected to find the magazines."

Jason's voice was weary. "Not to worry, Daniel. Do you have any conclusions yet?"

Dr. Ebo smiled. "Oh yes, we have tested both magazines and you were right all along. It is totally unknown, but as we know, a very deadly bacteria. Requires warm and moist conditions to thrive. Surprisingly easy to kill. Put it in regular air for ten minutes or pure oxygen for two seconds and it is ineffective..."

Jason cut in. "Is it something we know or something new?"

"We've no record of it. Your Cray was right however in that it's ancestors were definitely e-coli but it is a highly mutated strain. We have notified all law enforcement and sent messages to both publishers."

"We've done the same from here including the news services. Hopefully we can stop this thing before it goes any farther. Any antidote in mind yet?"

Ebo frowned. "No, Jason. The rapidity with which it attacks denies us the ability to retaliate in time. If you were there at the moment of inhalation I doubt if any efforts would be successful."

"Thanks a bunch, Daniel. I'll be in touch."

As Dr. Ebo's picture faded off the wall Jason said. "Cray!"

"Yes Mr. 'A'?"

"Get me Mr. Malawi please."

Malawi answered instantly and Jason brought him up to date. When he was done Malawi said, "We were too late."

"What do you mean?"

"Tru-Print has closed shop. There's nobody there and according to next door neighbors hasn't been for over a week. Not only that, but the building burnt to the ground night before last and arson is definitely suspected."

"These people are very adept at covering their tracks."

"Well," the SAIC[30] said, "They've had a long time to plan this down to the last detail so they have been miles ahead of us up to now. However I think, with your information, that lead is now shortened to just a couple leaps."

"How are your preparations for the meeting coming along?"

"We're practically ready."

"Has the money been raised yet?"

"Yes,"

"We need the banks where it is deposited."

"Why?" Malawi's tone betrayed his suspicions.

"We have to follow the money," said Jason. "If we have to wait until it is revealed to Luci we may not have time to set up to trace."

"That makes sense, Mr. Amador. Our money will be in the Federal Reserve Bank in Atlanta, Georgia and the other half of it will be in the Banco de Espana in Madrid, Spain."

"Good. We'll lock onto their accounts and see if we can tell you where it goes when it leaves there."

"Can you do it that fast?"

"We'll find out. While I've got you we've tracked a transmission that you should be aware of."

"Oh?"

"Yes, Satellite #YO422XX. They've broken off now but they've been loading it up with data for the last two hours..."

"What does it say"

"We haven't been able to break the code yet. When we break this connection I'll send you a piece of the data and you can set your computers on it."

"Any idea when they are going to transmit?"

"Undoubtedly during your meeting."

"Can you jam it?"

"Not unless we can get inside it. At this moment all we can do is track and record."

"Damn, these bastards have thought of everything..."

30Special Agent In Charge

"Hardly. We've taken their ace away from them. They can't kill the world with the bacteria any more -- that has got to hurt them. It may even force them to alter plans."

"Keep me posted."

"Ta." said Jason.

* * * * *

"Commander?"

"Yes, Number One."

"You missed the girl!"

"An oversight on Armand's part" The Commander said blandly.

"There will be no repeat of such stupidity, is that understood?"

The Commander's tone was deferential. "Yes Sir."

"They've constructed a special fire wall in front of the President. Double the amps in the first bolt -- I want no survivors."

"Done, do you know the banks yet?"

"No, they are known only to the Ambassador and the President. Oh, They have solved the riddle of the bacteria. It's not that important as it has outlived its usefulness anyway -- it won't change the plan. Time is tight here, if I don't get a chance to transmit again just stay with the plan -- we're almost there."

"I understand."

"I must ring off before they have a chance to trace this. One out."

* * * * *

"Mr. 'A'?"

"Yes Cray."

"I have all available information on the Island of Piros gleaned from The Encyclopedia Britannica, Compton's Compendium, The World..."

"Cray,"

"Yes, Mr. 'A'"

"I don't care where you got the information. The fact you've got it is good enough for me..."

"But there are twelve other sources...."

"Cray!, give me the goddamn information!   On second thought spare me the longitude and latitude."

"Yes, Mr. 'A'" Cray said and Jason could swear she retreated with hurt feelings. Christa smiled. Jason scowled at her as Cray continued.

Her delivery was in terse, concise sentences. " The island is situated twelve nautical miles from the coast of Turkey and is therefore claimed by both Greece and Turkey. Clear ownership has never been established. The island is a dark colored knot of rock eight hundred twenty two meters long north to south by maximum eighty meters wide and a minimum of 24 meters, east to west. It rises thirty meters high above the mean tide. Fifty foot high cliffs rise out of the Aegean on three sides but on the West side the cliffs are broken by a seventy five meter wide twenty degree sloping entry, cluttered up with house-sized rocks around which the sea crashes, making sea landings impossible..."

Jason interrupted her. "Cray, if you have an overhead shot of that area show it please."

Almost instantly a satellite shot of the island appeared on the wall screen.

"Cray, freeze it please."

The shot went static. It clearly showed the brooding island with marked shadows north to south.

"Zoom in please -- ten times."

The picture went blurry then refocused and the top of the island showed a series of peaks, bony backs and minor ridges with very little arable space and the only flat spot being utilized as a chopper pad.

"Zoom in please -- ten times." This time stone huts were apparent, partially built in, over and around the ridges and they appeared to become part of the rugged terrain. Rainwater showed in multiple cisterns.

"Cray, more information please."

Cray took up where she had left off. "It is recorded that in 1726 the Turks scaled the cliffs and occupied the rock probably as an advance guard post. In 1840 The Family Metaxis, a loose family of Aegean pirates stormed the island, killed off the Turks and have occupied it since. The Turks made an armed peace with them after the Turks had made two unsuccessful attempts to land and dislodge the Greeks. The island had one benign sea access which was destroyed by the present family in 1941 to keep the Germans from

114

landing. It was simply too expensive a project for the Germans and not strategically important so they left the island alone. Now it is only accessible by chopper..."

"...I'll bet that island is a rabbit warren underground," Jason said to no one in particular. "Cray, tell me about the Metaxis family."

"Little information is available. The Metaxis Family now has over two thousand direct descendants listed of the original islanders and nearly five thousand in other branches of the Metaxis family. They are spread all over the world. The highest reoccurrence of names within the family is that of A.M. Metaxis. That name appears on bank records, property documents, tax forms etc. This person appears to be the titular head of the family similar to a gypsy clan. Members of the family are engaged in many activities. One owns a small computer company, another owns a software programming company, a third owns a small bank in the Cayman Islands, a fourth owns a small shipping line, a fifth owns a taxi company, a sixth owns a small shipyard, a seventh owns a contract print shop and mailing company and an eighth owns several restaurants. A number are in jail or prison. Two hundred fifty are untraceable. Ten are tailors..."

"Cray, enough. Is there any apparent reason for the existence of those people on the island?"

"None."

"How many chopper flights have there been in the past six months?"

"There is a regularly scheduled monthly supply flight logged with the Athens Port Authority."

"Any way of checking how many unlogged flights?"

"Yes, by satellite reconnaissance."

"Cray, please check those reconn photos and get back to me."

* * * * *

CHAPTER 21
5 P.M. EST; 22:00 GMT; 24:00 ATHENS; 00 HRS

In the main assembly room at the United Nations the Secretary General of the UN, Robert Brownlee had deferred to the President of the United States. The President sat sternly erect in the center chair of the five behind a small table on a raised platform. At his request the table had been covered with a green velvet fire resistant cloth which hung to the floor on the camera side and concealed their legs and feet. The Vice President sat on his left and Brownlee on his right. To the left of Brownlee sat the portly Right Honorable M. D. Pochinski, United States Ambassador to the United Nations and on the far right of the President a burly Jed Foley visibly fidgeted, folding and unfolding his big knobby fists. In front of the quintet, and large enough to protect all of them, was propped a two inch thick panel of perfectly clear, tempered, bullet proof, laminated glass especially cast with nearly invisible, imbedded strands of foil. The best technical minds available to Malawi thought this might give the five a few seconds chance at getting away should Luci begin to throw bolts.

The area below and in front of them, the main floor, teemed with technicians of both genders while the press boxes lining the wall bulged with press people. The steady buzz of human babel hung in the room like the forced conversation at a viewing. Mingled among the crowd -- standing out like flyspecks in the sugar were security personnel from every available source. CIA, FBI, Secret Service, ATF, The UN Security police, New York's finest and a contingent from the New York National Guard. Each service had fielded a special SWAT team laden with the most advanced mobile weaponry known to man.

Malawi, as the security man of the hour, flowed through the mass of humanity with his entourage like a tall, dark mother shark dragging her babies through a school of unsuspecting tuna. Having been up all night, he was surly and short tempered and had spent most of the time stationed at the only door through which entry was allowed into the main chamber. He had demanded, and got, a senior man from each of the security detachments, and the work details, and together they had stood guard as each worker and security man reported for duty and was assigned his post. No one was allowed entry unless he was visually cleared by one of the senior security detail. No one was allowed off the floor once they were passed through the entrance. Restrooms were within the sealed area but anyone wanting to use them had first to clear the movement with their superior then sign in and out of the restroom through the National Guardsman stationed at the restroom door. All movements from one area to another on the floor were recorded by video camera and visual observation from the guards on the floor.

On the main floor of the assembly room the overstuffed chairs had been torn out just before dawn in order to avoid the possibility of missing covert electronics in them. Only since then had the area become covered with staggered tables and chairs, school room style. The total room had been scanned with every bit of science available to Malawi and all wiring had been summarily jerked out of the walls. There were now no wires in the walls nor the room except those the current crews had strung across the floor to handle lights, cameras, Mikes, monitors and recording equipment. Wires on the floor were like spilled spaghetti snaking in every conceivable direction. In various places in the room there were odd looking bits of equipment all seemingly trained on the five on the raised platform. These were SOTA scanning and sensing equipment designed to record the presence of Luci when Luci appeared. Along each wall was a long line of television monitors each of which would show a delegate who had elected to be present but only electronically.

The seats in the middle of the room now began to fill with the fifteen members of the Security Council brave enough to request attendance. At the door each was scrutinized minutely by a hand held metal detector as well as having to pass through the electronic arches. Once passed the rigid door check they were led past the armed guards, seated at a guard's discretion and given a set of headphones. They took their assigned chairs gingerly, warily watching those seated around them.

At nine fifty eight a.m. all were seated in hushed silence. At a signal from the floor director the wall monitors, one by one, brought a face from some distant place into focus -- some angry, some serious, some vacuous but all intensely frightened.

At the stroke of ten the Secretary raised his gavel, slammed it down on the table and intoned, "I now call this assembly to order."

Technicians froze at their assigned tasks, The Security members hunched their shoulders and waited expectantly and the buzz in the press booths dropped to zero.

The Secretary began the roll call and concluded with..."Let the minutes reflect that the fifteen members of the Security Council are in fact, physically present. Let the minutes reflect that the President of the United States and The Vice President of the United States are physically present as am I, Robert Brownlee, Secretary General of the United Nations; as is the Right Honorable M. D. Pochinski, United States Ambassador to the United Nations and Warden, Mr. Jed Foley. I now yield to the President of the United States."

* * * * *

"Mr. "A"!"

"Yes Cray," Jason had been watching the secretary on one of the wall screens. They were getting the yield from three cameras in the General Assembly room and each gave a different perspective. As Cray spoke Jason was already seeing the beginning of a text transmission up in the corner of one screen but he let Cray speak.

"Initial transmission to the satellite is in colloquial English. The word is 'Go'. The transmission is in process."

Jason looked at Christa and Ashton. "Thumbing their noses at us. They know by now we couldn't break their code so they want us to know it is about to happen."

"Cray."

"Yes Mr. "A".

"How long will that transmission continue?"

"At the present rate nine minutes and forty eight seconds."

Christa yelped, "Listen! There's the tone and there he comes!"

All eyes turned to the wall screens which gave three different views of the assembly room. Luci was beginning to materialize just in front of the raised platform.

Jason yelled. "Cray! Where did that word 'go' come from?"

"The Island of Piros, Mr. "A"."

"So that's the source of Luci," marveled Jason. He looked at the screen as Luci came into total being on the floor of the assembly. The tone was now a steady background hum, barely audible -- but there. "Now we know where your home is big boy. Let's see if we can disrupt your fun. Cray! Get me a strong copy of the Star Spangled Banner and put it on standby."

* * * * *

Malawi spoke softly into his Mike.. "I want every weapon dead on Luci. You've all seen his MO[31] on the video tape. If he raises his right arm and points it, do not wait for orders, fire!"

Lucifer stood and surveyed the room, turning to probe everyone with an insolent glare. Not a word had been said since he had appeared.

The President cleared his throat.

He broke the silence. "We are present as you requested, Mr. Luci."

His monstrous red head undulating like a giant snake, Luci snarled. His voice was so low that it seemed to begin inside you instead of coming from his mouth. "I have already taken note of who is here and who is not. I will deal with those who ignore me at a later time. In the mean time, let me conclude the business at hand. Who is supplying the money and where is it deposited?"

Pochinski coughed, and Luci turned the full brunt of his gaze on the Ambassador. The Ambassador's voice cracked but then gained strength and he said. "The United States is paying fifty billion and it is presently in the Federal Reserve Bank in Atlanta designated 'Luci'."

He stopped and cleared his throat.

In the dead calm of the room Luci stood with his hands on his hips, impatiently clicking those giant talons.

"Well?" he demanded.

Pochinski cleared his throat again. "The balance of the money has been raised by a consortium of European and Middle Eastern countries and is in the Banco de Espana in Madrid..."

31Modus Operandi -- method of operating or working.

"What are the code numbers?" Luci snarled.

Pochinski cleared his throat again. His voice trembled but he plowed on. "I, I'm not given the liberty of giving you those codes until you divulge where the money is going and what you are going to do with it.?"

"Egad!" said Jason.

"Shit!" Echoed Paul.

Luci was speechless. So was Christa. She, Jason and Paul watched the giant apparition stand there for long moments seemingly flabbergasted that Pochinski had dared voice the question.

Suddenly the Assembly room began to rumble. At first Jason thought it was the rare earth quake but it dawned quickly that it was the beginning of a cavernous laugh coming from the depths of Luci. Luci seemed to grow in size. He towered over the Ambassador. He spoke at leisure but spit his words at the Ambassador.

"You piddling pipsqueak! You see before you the embodiment of all that you have become and I am now your worst nightmare. Throughout the ages you humans, of all species on this planet, have enjoyed all the advantages of right and reason; intelligence, knowledge, understanding, compassion. But you have also destroyed it. One of your puny thinkers Lord Acton said it most appropriately "...absolute power corrupts absolutely."

He paused and glared around the room. No one moved nor spoke.

He went on, his voice a purr that made your toe nails ache. "In the millions of years since your conception in a bed of warm seaweed you have gained absolute power over your existence. You have had it within your grasp to nurture this world and exalt your kind to heights never imagined in the creation. Instead you have squandered that power by choosing to pollute your soil, defile your air and savage your neighbors. You took this bit of rock and raped it with liberal amounts of greed, avarice, and lust and you have left yourselves with nothing but a foul smelling ball covered with human shit."

He laughed and the building shook. "You invented laws, religion and morality to tame your masses and salve your conscience and almost from the moment of their conception you forsook them. In your cupidity you created your religions and in each of them you spawned me and now I am bringing you your own hell right here on earth. Numbers.." He rumbled, "... I don't need your primitive numbers. I am moving that money as I speak without your numbers and without your help,"

He turned away as if to depart then wheeled back to glare at Pochinski. His voice was derisive. "As for you Mr. Ambassador, I shall return when and where I please and those who displease me will die. For your insolence, I will let your successor know where and when to have the next installment of funds..."

Jason was to watch those next few seconds on tape many, many times and forever not to believe his own eyes.

Luci suddenly raised his muscular right arm.

The safety of every weapon trained on him came off and the fingers tightened on the trigger.

Luci threw a thunderous bolt, and all weapons fired almost as one. The prodigious fire power could be seen passing right through Luci. It was so great and concentrated that a mattress sized chunk of the floor was cut loose and fell to the floor below.

Luci's bolt struck the glass in front of Pochinski and paused, like a drill trying to force its way through a piece of seasoned oak. But in no more than the blink of an eye the glass turned a brilliant ice blue, shattered and the bolt thrust on. Pochinski was dead where he sat.

Luci was totally unaffected by the rain of death. He turned like a dancer spraying bolts into the pandemonium.

Through the bedlam, Jason, Christa and Paul watched spellbound at the slaughter transpiring on screen. Off to one side they could see secret service pushing the four live men off the stage and toward the back of the room. Others left standing in the smoke and haze started streaming for the doors.

"Run you fools," The apparition roared, "I've locked all the doors and you're all going to die!"

Jason leaped for a knob marked 'volume' and turned it full to the right.

Behind Luci's roar, and the snap and crackle of the bolts and the screams of the injured and the incessant gunfire still coming from troops left alive came the thunderous strains of "The Star Spangled Banner."

Luci's last act was to point his finger to throw a bolt but no bolt came and he faded from view.

"I've got you, you sunofabitch, I have got you," exulted Jason, looking up at the three screens now nearly obscured by smoke. But his bliss was short lived as the cries of the injured jarred him back to his senses. He said to Cray.

"Cray, get me Mr. Malawi if you can."

"What do you mean Jason, have you figured it out?" Christa queried.

"Wait until I get hold of Malawi," Jason said wearily, "and I'll explain it to all of you at the same time."

"Mr. "A", Mr. Malawi is on a scrambled line."

Jason pushed a switch and said with great effort. "Mr. Malawi, I owe you an apology."

"What for Mr. Amador?" Malawi's voice was heavy, the voice of a commander who has just come back from a battle he's lost.

"I've had the answer to this puzzle inside my head but I couldn't get at it. I must be getting old. First, let me ask you the bad news about your people."

Malawi paused and Jason thought he might cry but his voice was strong when he spoke. "Including Ambassador Pochinski we lost -- I lost twelve men and four women with twenty two wounded. "

'And the President?"

'Believe it or not we managed to save the President and the other three with just minor bruises. I had prepared a quick way out the back for this eventuality."

"I am most sincerely sorry, my friend but I think this may be at an end."

"What do you mean?"

"Metherell's theory."

"I don't understand."

"In the late sixties, Alexander F. Metherell[32] did some research and expounded the theory that holograms could be created with sound. The scientific world at that time scoffed and the idea was thrown on the trash heap..."

32In an article titled "Acoustical Holography" published on page 36 US International Abstract in October 1969, Alexander F. Metherell expounded: By 'illuminating" an object with pure tones of sound instead of with a beam of coherent light one can create acoustical holograms that become three -- dimensional pictures when viewed by laser light.

"Damn!"

"Yes and more. Evidently someone quietly picked up on the research and perfected it. Luci comes in on a beam of sound rather than a beam of light."

"Was it you that turned on the Star Spangled Banner?"

"Yes."

"Thank you." Said Malawi, relief in his voice. "If you hadn't popped in the National Anthem all my crew would have been dead plus all the technical people. He had us locked in and there was no way to get away from him. My tactical screen in front of the President blew up on the first bolt..."

"Yes, we saw that. Cray registered one hundred thousand volts at two thousand amps on that shot..."

"...Well, until you hit him with that overpowering music and broke him up we were all dead meat."

"Mr. "A","

"Hold on Mr. Malawi, yes, Cray?"

"The accounts at the Federal Reserve Bank in Atlanta and the Banco de Espana in Madrid have both been transferred..."

"Cray, Where?"

"...Banco de Isle in the Cayman Islands."

Jason turned to Christa. "Hack their accounts and see what happens to that money..."

"Ms. Christa programmed me to follow the money automatically." said Cray off handedly.

"Well?"

"Both accounts retransferred in the subsequent two minutes to receiving the transmission from Piros..."

"Cray, can you track where the money went?"

"Their computers went immediately offline -- there is no trace."

"Damn!. Malawi? You heard?"

"Yes. As I've been listening I've had myself patched through to Satellite control. They tell me there is an eye orbiting over the Caymans and they may be able to give us some information. Hold on a moment..."

Jason, Christa and Paul fidgeted for a few moments until Malawi came back on the line. "Mr. "A", you there?

"Yes."

"G.P.S. places that bank at the corner of Dominguez and Pinata. The building they were in and half the city block exploded right after the transmissions ended -- everything is gone."

"Dead end," said Paul

"Yep," said Jason. "That money is now in the hands of the enemy and we know not where -- nor will we ever."

"I'll keep you posted," Malawi said and rang off.

"Cray," said Jason.

"Yes Mr. 'A',"

"Get me all the information you can find on the Banco de Isle in the Cayman's. Particularly I want to know who owns it -- and wake Mr. Chaney -- we've got to get him started for Piros."

* * * *

CHAPTER 22

"Armand!" The tall man's deep voice demanded in his spaghetti mike, "Close out and come in here." The tall Commander was ensconced in a swivel chair in front of a low console which backed against the large tinted window spanning the length of one side of the control room. The window viewed the main cavern which held the three rotating Luci cages. Each was occupied by a Cyber Pilot. To The Commander's right and left were three technicians, each backing up a Cyber Pilot in the cage. The Commander's position at the console combined the controls for all three cages so he could keep track and/or alter anything that was happening in any of the cages.

Out in the main cavern one of the giant round cages came to a grinding halt. Armand could be seen inside, suspended in an upright position, clad in a skin tight suit, dozens of sensors attached to his body. On closer inspection the ball was a simple frame of steel and pipe with holes for high pressure ejection of air toward the center. The intense equalized air pressure suspended the pilot within the ball but allowed him to move in any direction. The sensors provided a constant two way flow of haptic[33] information to a thousand receptors set inside the steel ball. A miniature panel attached to his helmet in front of his head rotated with him and he controlled the movements of Luci with his own body. The horrible bolts were dispensed by the simple feat of pointing the right arm. This was especially wired for the effort.

Armand reached up and toggled two switches on his panel and the hissing diminished. As it reduced to nothing he could be seen to splay his feet and catch himself on built in supports on the inside of the ball. He disconnected the loom of sensors from his upper and lower torso then grabbed a bar next to the opening and swung easily

---

[33]"Haptic Interface" -- Anyplace mechanical energy takes place between the body and its environment.

through a door in the frame of the ball to a small platform. He pulled his helmet off and came striding into the control room.

Expecting praise for a job well done he was surprised to be met with instant inquisition. "You were losing it right there at the last. Why?"

He stiffened. "It was the American Anthem, Commander." He almost accused the Commander of being lax in not excluding extra music from the scene of his triumph but thought better of it and continued somewhat contritely. "We are supposed to have all extraneous sound blocked out during Luci's sessions but someone started piping the Star Spangled Banner in...?"

"Couldn't you override it?" The Commander interrupted sharply.

Armand ran his fingers through his hair and carefully pointed out the obvious. "Not without trashing everyone's brains including Number One."

The Commander was pissed and not prone to discussing opposing viewpoints with lowly pilots but he restrained his ire somewhat and replied, "Well...you came out in time to keep the crowd from recognizing a problem but someone has definitely caught on to us."

Armand ventured an observation. "Do you think the Feds?"

"No, they're four steps behind us. It has to be Amador." He leaned back against the cool, dark wall and scratched his short beard. "Go on and get a shower and get some sleep -- I'll need you before your next shift."

"Yes sir."

The Commander glanced at the other two pilots who were busy injecting Luci into other unrelated scenarios. Damn him, he thought, we should have been able to wrap this all up and have done with it without anyone knowing the truth. Now we'll have to kill him -- or at least neutralize him until it's all over.

"Commander," a technician pointed to his console, "#1 is on the scrambler."

He shrugged and punched a button on his console. "#2 here."

"Well it was chaos. We managed to kill most of the people in the assembly room but Armand missed the President and his group except for Pochinski..."

"Are you alright?"

126

"I'm tired as hell but I'm alright -- Malawi got us out the back. How about the transfers?"

"All of the money has been retransferred and the records obliterated."

"On schedule?" asked #1.

"Yes, it all happened in less than two minutes."

"Good, but we've got some more problems. Never in a New York minute did I figure anybody catching on to our program but that goddamned Amador has a photographic memory and he's now told Malawi about Metherell's theory. They've got us cold on the sound and I've got no plan 'B'...."

"Yes we do."

"What are you talking about?"

The Commander paused and savored the moment long and wide with a huge smile on his face. Finally he said. "It helps to hire the best scientists money can buy..."

"Hurry up!" snapped #1, "I can't keep this line open much longer."

"The researchers have discovered a way to combine ambient light at the destination to strengthen the hologram. They've also figured out a way to keep the original line open so it can't be shut down..."

"Then you mean we can keep Luci going and make a second collection. Is it operational?"

"Within the week."

"Good." At that moment one of the Commander's technicians laid a note in front of him. "Hold on," he said, reading it quickly. Then he resumed. "#1 -- the American is buying scuba gear and arranging for a bird. Looks like he is going to visit us very soon."

"Good, put him out of his misery after you get him to tell you all about Amador's plans. The pilot is ours is he not?"

"Yes, in fact he's my son in law Enrique

"Good, handle Chaney however you want. I'll handle Amador personally -- I've got to go. Out."

* * * * *

127

CHAPTER 23
8 P.M. EST; 0100 GMT; 0300 ATHENS

Peter quietly noted the considerable skill of the chopper pilot as he coaxed the bird into the air and pointed her nose to the east over the slumbering city of Glifada on their way to the Aegean Sea. He was young, perhaps twenty five, wind worn like most of the Mediterranean people and mustachioed, and appeared to be very much the young swashbuckler out to impress his strange American fare. He had invited Peter to sit in the second seat and now pointed to various lights below them and gave him a running commentary in broken English. He had not seemed perturbed by this spur of the moment crazy 'Americain' who suddenly appeared out of nowhere and wanted to go to a tiny island just off the coast of Turkey in the middle of the night yet.

In truth, Jason had cautioned Peter to expect that Metaxis would be aware of his chopper ride and even the possibility the pilot might be in their employ.

Scuba gear had been the most difficult to acquire at 2 a.m. but the pilot had been most helpful and for an extra five thousand drachmas had routed a friend out of bed and the result now lay in a careful heap behind Peter's seat.

For his part the young pilot was ecstatic. As soon as the American had made contact he had called the island for instructions. He had spoken to the Commander, his father in law and received his personal instructions about what to do with this Yankee. He was to drop him off in the water wherever requested and the islanders would be waiting.

But a recurrent thought kept scurrying back and forth in his mind. What a triumphant entrance it would be to land the chopper on the pad and dump the American out bound hand and foot. The more he thought this, the more attractive it became. As the chopper broke loose from the land and soared out over the cobalt Aegean following the path of the yellow moon he could think of nothing else except when would be the best time to whip out his pistol and subdue his fare.

* * * * *

"Admiral, I want every damned one of these bastards alive so they can stand trial -- if at all possible."

The President of the United States gazed out the window of the oval office at a flock of birds sleeping fitfully in the lighted cherry trees. He turned back to the Admiral.

Honored with a chest full of medals and ribbons by his country, lauded by his colleagues, respected by his enemies and his peers, tall, fit, Admiral Dennison J. Kincaid stood respectfully at ease as his Commander in Chief ranted and fumed. He had been summoned to a one-on-one with the Chief at this unusual evening hour. In his hands were a number of rolled letter sized sheets of paper which had been thrust into his hands by an aid as he darted into the President's office..

The President had taken an hour of elective solitude after the debacle at the United Nations Building. It had taken him most of that time to stop the shaking in his hands and the tremors in the rest of his body that were the aftermath of the awful three minutes in the Assembly room. Then he spent the remainder of the day in frantic meetings with what was left of his Cabinet. Pochinski had been drilled neatly through the head with a force so hot and awesome that his brain had turned into solid rubber and the hole was cauterized all the way through to the back of his head. He and the rest of the wounded and killed had been cleared out of the room and it had been closed and made off limits to anyone not personally admitted by Malawi. Malawi had then spoken to the President about Jason's suspicions about Piros. The President had then spoken personally with Jason about Piros and the subject of this meeting with the Admiral was to plan and launch a clandestine strike against the island by chopper using Navy Seals. Hence the presence of the Admiral.

The President sat down at his desk and looked up at the Admiral. "Have you had time to bring the pot to boil?"

"Yes Sir. It's a relatively simple attack. With your permission..." He spread the maps on the desk in front of the President and pointed at Greece. "...It's less than an hour over there so we'll take off from the old Hellinikon Air Force Base an hour before sunset in four Super

Stallions[34], each with twenty five Seals. We'll send two Super Cobras[35] along to ride shotgun. We'll send all six of those birds straight in out of the setting sun. I've looked at every satellite shot we've got of that island and I can't see any weapons of any kind, at least on the surface. Considering it's size and who owns it one hundred men plus the heavy weapons on the Hueys should be plenty to take care of anything there. However, we're going to also send the four Hueys slightly to the north, let them pass the island then drop south and come back in at water level. This strafing diversion will be timed to hit the island three minutes before the Stallions "

"How soon can you launch?"

"An hour before night fall tomorrow, Greek time," Kincaid assured him. "Just a matter of assembling the personnel from various bases. The supplies and birds are already on Hellinikon."

The President tepeed his hands and blew through his fingers. "Admiral, I want it understood that we are going in there to capture for interrogation purposes only. This is not search and destroy. There's a lot we can learn on that island. " He skewered the Admiral with his intense brown eyes and waited.

The Admiral stepped back and came to attention. "Understood Mr. President."

The President smiled. "Relax Dennis, it's just that the world's eyes are upon us and we can't afford to screw this up -- understood?"

Kincaid smiled, they had been friends for a long time. "Yes Sir."

"Good, have the orders drawn up and I'll sign them tomorrow evening just before you embark. Have you picked a leader yet?"

"Yes, a young Commander who cut his teeth in Vietnam."

"Do I know him?"

"Yes sir, you recommended him for a Navy Cross when you were Secretary of Defense. Commander Joshua Templeton."

The President flopped his hand in the air -- remembering. "That's that young -- God he was just a Lieutenant then -- that young Seal

---

34Sikorsky CH -- 53E Super Stallion. 99 feet long with a 76 foot rotor diameter will cruise at 173 mph carrying up to 55 troops.

35Bell AH -- 1 Huey Cobra. Twin -- engined Marine attack helicopter which saw prominent use in Desert Storm. Carries pilot and co -- pilot and maximum armament..

who dragged four of his wounded buds out of that chopper when it was under twenty feet of water..."

"The same..."

"God! That was a lifetime ago. Saved 'em all, if I remember."

"Yes Sir."

"So he's a full Commander now? You think he's got the balls for this operation?"

"Yes Sir."

"A lot's depending on this. You'd better be right."

"I'm confident."

"Ok. For his ears only, notify him to pick his men and get ready. Fly in whoever he wants." He glanced down at his desk then back up at the Admiral.

"What have you got to tell me about the submarine idea?"

The Admiral fidgeted. "We haven't got a thing in the Mediterranean and nothing even relatively close. The best we can do is use a Spanish diesel which is on maneuvers near Crete..." He consulted a sheaf of notes in his hand apologetically. "Mr. President, I was handed these notes just before I walked in here." He looked closer at them. "Seems there is an Italian Diesel Boat, owned by the Spaniards, of the..." He read straight from his notes... "Salvatore Pelosi class, the Hernando DeSoto, currently on shore leave in Iraklion, the Capital of Crete..."

"God, what a lousy place for R&R..."[36]

"Yeah, isn't it the truth, but the whole crew is made up of kids on their first submarine cruise. There's only three experienced men aboard. The Captain, his second officer, a Lieutenant and The Chief of the Boat. But they could be there in about twenty hours under full steam -- providing we can get them to participate."

The President snorted. "They won't have any choice. How big is that boat?"

"Two hundred eleven feet."

"Jesus, a boatload of kids in a thermos bottle. I didn't think anybody still had anything that small. D'you think we can stuff twenty SEALS aboard?"

36A military term -- -- Rest & Relaxation.

The Admiral shrugged. "In this kind of an emergency -- yes. They'll have to 'hot bunk'."[37]

The President gazed back out the window. "They shouldn't be aboard long enough to worry about that. I want personal interception of that boat. No radio contact. Get the SEALS and enough explosive to sink that island aboard the sub and give that Captain orders to submerge and drop..." Here he turned to look at a map of the Mediterranean sea floor. He stabbed his finger at it. "Into the mouth of the Edremit Trough -- right there. They are to hover there dead silent at periscope depth until we either give them the order to go in or we call it off. "

"Aye aye, Sir"

"Dennis, I want this kept between me and you. Everybody else is 'need to know'. You take orders only from me and they take orders only from you. If anybody gives you shit you lock 'em up till this is over. Understood?"

"Yes sir."

"D'you speak Spanish?"

"No sir, but my aide does."

"Good. When you leave here I'm going to put through a call to the Spanish King to clear the way but if that Spanish Captain misses the big picture haul him and his crew off there and put one of ours on. It doesn't make any difference who's crewing it as long as it's standing off Piros twenty hours from now. Keep 'em on tap there until we give the choppers a try. "

The President jumped to his feet and grinned. "Listen, old friend. You'd better take fifteen minutes in the other room and put this on paper. Two copies. I'll sign both of them and keep one. That'll get you off the hook if something goes wrong."

The Admiral breathed a sight of relief. "Yes, sir."

---

[37]In the smaller submarines frequently there are more enlisted men aboard than there are sleeping accommodations. Consequently three men may be assigned to just two bunks or one bunk. While One man is on watch and another man is up, around and being trained or otherwise occupied the third man sleeps. ('Hot bunk' -- the bunk is never empty.)

CHAPTER 24

"Hello Peter. I Have some interesting news for you."

"Yes Mr. 'A'?"

Jason's voice in the cell phone was quiet, reassuring. Almost too quiet to be heard over the whine of the chopper's turbine. Peter turned away from the pilot the better to hear -- and not be overheard.

"Don't let your expression relay what I'm going to tell you. OK? Cray tells us the Metaxis family owns the chopper service you hired so they know exactly who you are and where you're going..."

As the words sunk in Peter turned nonchalantly back toward the pilot to find himself staring into the dark, round muzzle of an automatic pistol. It was just visible in the pale moonlight. The bore looked big enough to stick his thumb in. The pilot motioned sharply for him to put the phone down and his hands up. Peter put one hand in the air and with the other very quickly let the phone drop into his side pocket.

* * * * *

"Cray!"

"Increase the volume on Mr. Chaney's cell phone to max."

Seeing the look of concern on Jason's face Christa said, "What's up?"

Jason held his hand up for silence and donned a pair of headphones. He pointed then to Christa and Paul. Each quickly put on a pair.

What they were hearing above the whine of the rotors was definitely not the sound of a Sunday afternoon coffee klatch. Several loud thumps, suddenly two shots, a scream and a groan, followed by the banshee wail of a helicopter in full drop. A hard whump -- then -- silence.

Jason looked up at the screen. Cray put the last known position of Peter's chopper at one mile due west of Piros.

"Cray! Voice analysis of that scream -- was it Mr. Chaney?"

"Negative."

"One small tic in our favor." Jason said gravely.

"Mr. 'A', I have two bits of requested information." Said Cray unconcerned.

"Cray, go." Jason said eagerly, laying down his headphones, expecting some word on Peter. .

"Reconn photos from satellite show there have been twenty four unlogged flights to Piros in the past six months. Fifteen of those were cargo carriers -- the remainder carried passengers. Crates unloaded appeared to have mass approximating that of heavy machinery..."

"Or weapons..." interjected Paul.

Cray continued. "And The Banco de Isle is a Caymans Corporation held as a blind trust by a French consortium named "Tropico Internationale..."

"Oh damn the Island Bank," stormed Christa looking up at the screen. "Cray, tell us where Peter is."

"Mr. Chaney has separated himself from the remains of the chopper," Cray answered calmly, "and is proceeding towards Piros..."

"WHAT!" the three said as one.

"His cell phone is deactivated but I am tracking his G.P.S. pulses and his physiology is normal except for his diastolic pressure which is forty points above..."

"Cray, why didn't you say something?" snarled Christa.

"I had no query." retorted Cray, his tone infuriatingly self righteous.

Jason snorted, his mind already on something else.

"Christa, do you still have that worm[38] you wrote that steals information then wipes the hard disk?"

Christa grinned. "Yes."

"Think you can get it into Piros."

"Probably. What do you want it to do once it gets in."

"I want a complete download of everything on their hard drive then I want to wipe it clean. That'll put them out of business and give us the technology."

---

[38] A computer bacteria that burrows into your data and waits for a predetermined 'trigger' to do whatever it is programmed to do.

ONE HOUR LATER

"The American is a good swimmer."

"Yes, pity."

"He is rounding the northeast corner now Sir."

"Good. Are we ready for him?"

"Yes Sir. As soon as he comes through the gate the squad will have him."

"Use the net."

"Yes Sir."

* * * * *

Peter cruised comfortably at ten meters below the surface until he was within one hundred meters of the island. There was no entry except on the east, northeast corner where the slope came down to the water so he made a wide left circle calculated to bring him even with the base of the slope. The nearly one mile swim in the calm blue water beneath the surface had allowed him time to regulate his breathing and his heart rate. The fight with the chopper pilot had been short, brutal and unexpected and had been over almost as soon as it had begun. Jason's words had given him the one moment's edge necessary for surprise. Striking out at the pistol was a trained, unconscious reflex when he turned and saw the weapon. In the flash of an eye that it took his adversary to realize that his prey was attacking Peter only managed to deflect the mouth of the weapon away from himself. But the two shots the pilot got off brought the bird down. On the way down his 'misjudgement' of this 'Americain' cost him his life. Peter made it back into the cushioned seat just before impact with all sorts of escape scenarios racing through his head. The chopper dove straight into the dark blue water and began to sink. The impact shook him sorely but he escaped with minor bruises and the cool water rushing into the fractured bubble cleared his senses. Even so he was forty feet under water and dropping for the five hundred meter bottom like a stone before he located his scuba gear and fought his way clear of the descending wreckage in the dark. In the confusion he lost the Locator and his Polymer Pistol.

He let his head break the surface a hundred meters from the slope and backed water while surveying the territory. The gloomy bulk of the island stood thirty meters high and faded out into the night to the left and right of him. It was a bleak, forbidding piece of rock broken only by the slope surprisingly lighted from above.. The slope had been created in 1941 by several well placed shaped charges[39] that had let a set of steps up the face of the cliff slide into the sea and left the slope rumpled along fault lines and strewn with rocks some taller than a man. The slope did not quite reach the sea but sat there like a monster Easter Island Monolith staring morosely down at him. The base actually started almost two feet above the reach of the tide. A narrow, choppy pathway led up the left side of the slope dodging in and around the rocks. Five meters up the slope a chain link fence crossed the slope parallel with the tide and stopped any casual passerby from striding up onto the island. The fence line was broken by a gate which was shut.

A stiff breeze was kicking up and he watched as large phosphorescent waves rolled harshly up to the bulwark, slapping it, sliding over, then slithering noisily up the slope past the fence. The froth banged in and around the rocks perched on the slope and finally ran out of strength just above the fence. The slope and the rocks were polished by the ceaseless energy of the salty sea and glistened like black ice. He could see where the slope edged onto the top of the island just below the lights. Beyond that the island met the brooding ebony sky thirty meters above his head. Unfortunately for anyone approaching from the sea the floodlights were all on the top shining down so any trespasser had to walk straight up into the glare. He shrugged in the water -- nothing was ever easy.

He moved closer to the wall and found there was a series of rough cut steps in the rock leading out of the water and up the two feet onto the slope. Nothing was moving up the slope so he kicked off his fins and dropped his rebreather[40]. When the next wave made its upsweep he popped up the steps, rolled out onto the harsh rock and let the waves help him scrabble up the slope to the base of the gate. He lay there, breathing easily aware that he was totally exposed but no

---

39Shaped Charges -- An explosive charge that can be directed in one direction.

40'Rebreather' --   -- a diving device that recycles a diver's own breath. Extremely useful in 'stealth' situations because it does not leave a trail of bubbles to identify the presence of the diver.

help for it. The hackles raised on the back of his neck as he examined the gate and found it unlocked. Come into my parlor, said the spider to the fly, he thought as he opened the gate and slid through in one fluid motion.

* * * * *

They were standing next to the White House helicopter pad. The White House yard lights were blazing in the mid evening darkness accentuating the shadows playing around the group. The President of the United States extended his hand.

"Jed, there is really no reason for you to stay on here in Washington..." With a warm smile the Warden took the hand and nodded his gargoyle head. "...You have been of unparalleled help to me and to the nation in this time of extreme crisis."

"Thank you Mr. President."

"The United States and I are forever in debt to your courage. If we can ever repay that debt you have but to ask". He indicated the waiting helicopter. "I've given my pilot instructions to take you back to your chopper pad at Attica so that you can rejoin your wife. I am sure she has been more than anxious these past few days."

"The Gorilla" as he was affectionately called by those close to him seemed reluctant to leave. He turned and faced the President who was flanked by Secret Service men and the ever present Malawi standing stiffly off to one side, carefully monitoring the situation. Foley glanced around at the entourage and his huge face split into a tired grin.

"It's been an interesting time, Mr. President, I certainly hope my next trip to Washington can be under more benign circumstances. What happens now?"

"The Chiefs of Staff are working on a number of plans, including sending a sub to attack the island." The President answered. "I'm sure they will come up with an answer to this menace". He turned half way away from Foley as though anxious to get back to Affairs of State. He was obviously bringing the conversation to an end. He waved Foley toward the chopper. "Bon Voyage, Warden," he said, then turned wearily back toward the White House.

* * * * *

Peter scrambled up the misty pathway against the lights. Halfway to the top he stopped as a rough dressed peasant type stepped from behind the rocks, blocking his way. The man wore a cap pulled low over his eyes, shadowing his bearded face. With the light behind him he was simply a large man shaped blob above Peter. Peter advanced on guard but heard stealthy movement behind him. He turned swiftly, relieved to have the light at his back Just out of reach below him the twin to the assailant above him stood stolidly with a sturdy shepherd's staff at his side. It was then from behind that Peter felt more than heard the soft whoosh of the weighted seine. It caressed his head and shoulders and formed a supple, impenetrable cocoon around him. He saw the staff rise in the air and tried to raise his arms to shelter his head but was powerless to dodge the blows.

* * * * *

"Mr. 'A'?"

"Yes, Cray."

"I am still holding the remainder of the information on the Banco de Isle."

"Cray, Give it to me."

"Do you want that or the latest information on Mr. Chaney?"

Christa and Paul both turned around from their stations. "What's happening to Mr. Chaney," Jason asked.

"Inconclusive, " answered Cray. "He had some distinct trauma ten minutes ago but now the readings are indistinct and inconclusive. There is interference with the signal."

"Can you tell if he's alive?" Christa asked anxiously.

"Negative," said Cray. "G.P.S. is now only giving me his location which is near the middle of the island of Piros and approximately five meters below sea level..."

"They've got him underground." Paul said

"So he's apparently been captured," Jason added staring hard at the screen as though he could read something extra from the terse data there. He looked at Christa's teary eyes. "We expected this possibility, Christa," He said softly. "Peter knows what he's doing. If anybody can make it happen it will be him."

He turned back to Cray. "Cray, what else do you have on Banco de Isle?"

"The French Consortium which owns it is composed of seventy two owners, sixty three of which are related to a Metaxis or carry Metaxis surnames..."

"Metaxis again!" snorted Jason, drumming his fingers. "I'm beginning to think this whole thing is simply about money..."

"Surely they wouldn't kill fifty thousand odd people *just* for the money?" said Paul.

"Man kills on the streets every night for just enough to buy a rock, Paul. One hundred billion dollars is a whole lot of money and that's not the end of it. They're coming back for more unless we stop them."

He looked up at the clock. "Cray, get me Mr. Malawi."

It was but a moment before Malawi answered.

"Don't you ever sleep?" Jason chided.

Malawi's voice was tired and quiet. "My inner circle of friends seems bent on giving me information at these odd times -- what's up Mr. 'A'?"

Jason was apologetic. "I'm sorry but this can't wait. You must get word to the proper people that we have an injured friendly on Piros..."

Malawi was instantly alert. "What? -- who?" Then he said, "It has to be Chaney."

"That's right, my right hand, Peter Chaney. Here's how it came about..." Jason said and spent the next five minutes giving the FBI Agent all the information he would need when he informed The President.

When Jason finished he asked, "What's the latest on the battle plan.

Malawi filled him in and finished by adding..."This action is totally off the wall and fluid at this moment and this enemy is so totally foreign to anything we've met before that no one and I mean no one is venturing even a guesstimate on the outcome." He stopped, drew a deep breath and added lamely, "All we can do is do it and hope."

Jason agreed then he yawned. "I think it's time we all got some sleep. We'll talk more later."

He turned to Christa and Paul, his face mirroring his age. "I don't know about you two but I'm exhausted. We all need rest but I do think someone aught to be keeping Cray company in case she comes up with something new on Peter. Paul, you and Christa flip a coin about who stands the first watch. I'll be up and around at about five so I'll relieve whoever has the number two."

* * * * *

"#2. We have another problem developing. The US is deploying a WWII diesel submarine into the Mediterranean. It's mission will be to slip in and quietly capture Piros..."

"Why not use a nuker?"

"They don't want to destroy our technology but the real reason is because nobody has a nuker in the Mediterranean. Besides if they just wanted to eliminate Piros they'd just sit out in the Atlantic and do it with Tomahawks."

"Whose boat is it?"

"Spanish operating under the mantle of NATO."

"How soon?" #2 asked anxiously.

"ETA Piros tomorrow evening. But you shouldn't worry."

"Why not?"

"To get near the island they have to come up the Edremit Trough[41]. When they get close send Armand aboard the sub and swamp it. You can drop them at the bottom of the trough off the island and nobody will ever find them under five hundred meters of beautiful, blue Mediterranean water..."

* * * *

---

41 Edremit Trough: One of the many underwater valleys breaking up the floor of the central Aegean Sea.

140

CHAPTER 25

Ashton drew the first watch. The many hours of crisis had weathered him and made him comfortable with Cray as well as Jason and Christa. Although he was finding it exceedingly more difficult to keep his mind on his tasks when the lady was present he more dreaded the time when Chaney would make it back. Part of him secretly hoped Peter would fall prey to *something*. It would be great if he would just go back to California when this was all over and not come back. Paul was sure in his heart that, even though Crista protested, she still had a serious underlying interest in Peter Chaney -- and dammit, he couldn't help but be jealous.

In slack times Christa had shown him the way to the games on the Internet so he tidied up first then sat down for a quick review of the headlines then decided to enjoy some light entertainment. Without thinking, he flicked a dial on the console and brought in some soft, modern Tchaikovsky.

* * * * *

Lieutenant Commander Santiago Arredono Espinoza stood uneasily within the bridge of the Hernando DeSoto. Tall, slender, tanned, mustachioed, blond and blue eyed he was the epitome of the Castillian swashbuckler from whence came the finest swordsman in Spain's early history.

He watched with great interest as the huge American Navy Sikorsky chopper hovered cautiously over the dock next to his submarine and finally dropped down to the tarmac where it bounced warily on its haunches. Even fifty yards away the wind from its slicing blade blew him about -- nearly blowing his hat off his head. He didn't care. He felt his zenith rising. What a blow it had been to be ordered to take this moth eaten old boat full of raw recruits out into the middle of the Mediterranean and teach them to be submariners when his colleagues were lounging around the court and enjoying the ladies. But with the arrival of that pregnant grasshopper sitting out there on the dock he could feel the worm turning so to speak and he was ready for it. Sometimes you lucked out and found yourself in the right place at the right time. And this, he felt, was the beginning of his moment.

He was still reeling from the scrambled phone call. The King no less -- his king – the King of Spain, had called him personally. Had got hold of him on this boat, had literally drug him out of bed at slightly before four a.m. and had electrified him by calling him by his first name and asking him, for the glory of Spain, to cooperate fully with the Americans who were landing in this helicopter. Any thing they wanted, no matter how odd up to and including putting his boat under their command and taking them where and how far they wanted to go. The honor of Spain rested squarely on his shoulders. As the chopper settled down He drew his tunic more tightly around his considerable shoulders and slid over the side of the conning tower and down the steel ladder to the deck. He was followed by his officers and the Chief Bosun. They all turned to watch the chopper.

The rotors began to slack off and a side door opened. First man out in the late morning darkness was a small, compact figure in brisk navy khaki. He looked about for a moment then stepped quickly away from the ship towards the gangway leading up to Espinoza. He was followed by a large young man in the full camo[42] battle gear of the American Navy SEAL. Both stopped at the edge of the boat and saluted him then continued across the gangway. Espinoza involuntarily stiffened as he caught the glint of the Full Commander's silver on the stranger's collar. In his eagerness he had to remind himself it was unwritten protocol for the boarding officer to salute first even though there might be a slight difference in rank.

The U. S. Navy Commander was shorter than Espinoza by six inches but what he lacked in size was overcome by his presence. He seemed larger and larger as he came up to the submarine Captain. He stopped in front of Espinoza, saluted smartly again and asked, "El Capitan Espinoza?"

Espinoza returned the salute and said "Si."

"Permiso abordar, por favor?" The Commander asked in halting Spanish.

---

42 Camouflage

Espinoza smiled and said, "Bienvenido a bardo."

The American held out his hand to shake Espinoza's and spoke at the same time. "Joshua Templeton, U. S. Navy. Begging your pardon Capitan, are you comfortable with English?"

Espinoza grinned -- all that laborious grade school work would now come in handy. "Yes, Commander.."

Templeton looked up at him, relieved, and returned the smile. He looked around, took note of Espinoza's second and third in command and the Chief, nodded to them and said. "Good, that will make our task today much less difficult." He referred to the large young man who stood slightly behind him. "This is First Lieutenant Charles Hopkins." He pointed to the ranking Lieutenant at Espinoza's side. "Is this your second in command?"

"Yes, This is Lieutenant Antonio Rodriguez"

Templeton shook hands with him and one other junior officer who stood in the background. He turned back to Espinoza. "Capitan, Is the Chief of your Boat present?"

Espinoza indicated the Chief who saluted smartly. "Yes, Chief of the Boat, Carlos Costanza."

Templeton returned his salute. He spoke to Espinoza while indicating the Chief and the Second Officer. "May we go below where the five of us can talk?"

Espinoza told the junior officers to retire to the bridge and led Templeton and Hopkins down the hatch. The way led along a cramped passageway where it was necessary for everyone but Templeton to duck away from gear projecting into the right of way. Rodriguez and the Chief followed them into the Captain's narrow stateroom. It was barely large enough for all five men to press in around the small work table in the center. A single made-up bunk lined one wall and served as the chair for the table. At the end of the room a narrow door stood ajar revealing a commode and a shower. Templeton looked around for a moment then shut the passageway door.

He turned to Espinoza and wasted no words. "Your people have called you?"

"Yes my king spoke with me this morning."

"You are ready to help?"

Espinoza answered without hesitation. "Yes. Without question -- whatever you need."

Templeton looked directly at Espinoza. "Captain, what I have to say is for this room only. Not even your third in command is to know what you are doing. Understood?"

Espinoza nodded his head gravely. This man was echoing his King's words.

Templeton asked. "Do you have a topographical map of the Mediterranean sea floor?"

Espinoza motioned at Rodriguez who tilted a panel away from the wall, brought out a map and spread it on the small table. Templeton pointed to a spot just west of the island of Piros in the Edremit Trough. "How soon can you put your boat right there?"

Espinoza glanced up at a wall clock which stood at five fifteen a.m. He shrugged, "An hour or two to collect my crew ashore then at flank speed I can be there at four p.m. this afternoon."

"I've got twenty SEALS and their ordnance out there in that chopper. They've got three day's rations and six DPVs[43] with them. These particular DPVs are state-of-the-art and are designed to pull up to six men. But they're kinda bulky and designed to be carried in the torpedo tubes. Can you work them in, knowing that this shouldn't last more than forty eight hours?"

Espinoza looked at his Chief who nodded his head. "We will make the room. What do I do when I get there?"

Templeton was grim. "In case Plan 'A' doesn't work you are our very important Plan 'B'. Piros is the world headquarters for Luci..."

Espinoza caught his breath and exchanged glances with Rodriguez. Templeton continued..."For your ears only we are making a chopper assault on that island late this afternoon. If something should happen that it doesn't work out you will lay my SEALS off so we can make a sea assault."

"How will I know?"

Templeton was terse. "You will string out a wire antennae on a float, drop as far below it as you can, lay dead silent in the water, and

---

43 Diver Propulsion Vehicles. Underwater, torpedo- like, battery operated machines designed to pull a diver rapidly through the water.

just wait for word. You will receive orders to go in or come out depending on the outcome of the original assault."

Capitan Espinoza thought for a moment then turned abruptly to Rodriguez. As a courtesy he spoke in English. "Take Lieutenant Hopkins up and bring the SEALS aboard while it's still twilight."

He turned to Chief Costanza. "Put them aft in the central passageway. Show them the head. Make sure they have plenty of water. Send someone into town to spread the word that we will be putting to sea at eight a.m. and everyone is to be back aboard immediately -- but do not tell them why." As an afterthought he said, "Get those DPVs inside the tubes."

Templeton smiled, he liked efficiency. Espinoza turned back to him. "Anything else, Commander?"

Templeton heaved a sigh of relief, this had been far easier than he had dreamed. He shook Espinoza's hand. "No Capitan. Lieutenant Hopkins has other information for you but as soon as the chopper is unloaded I'll be off -- God speed to both of us."

* * * * *

Luis Martinez was sitting comfortably at one of the sidewalk tables scattered at the edge of the crowded Central Square in front of the Taverna Iraklion in Crete's capital city, Iraklion. Even though very early, the rising sun was brilliant and warm and, under his umbrella, Luis sipped his cappuccino and was at peace with the world. He was not a large man nor was he important looking. In fact if you were introduced to him you'd probably forget his name before you had time to turn around and if you saw him in a crowd, dressed in his seaman's togs and wanted to know something Luis was the last man you would ask. He was one of those myriad little people who make up 'the crowd' wherever people gather to socialize and pass some time. So he was somewhat surprised and annoyed when the swarthy man in the expensive suit materialized out of the early morning crowd and sat down across from him. The stranger said conversationally, "You look like a submariner."

Luis wasn't sure the swarthy one was in fact talking to him except that he was sitting at his table and he was looking right at him and he was speaking perfect Spanish. If you sat very quiet in the Square and you really listened to the din around you, you could hear every major language spoken in the world, including Spanish, but most of Luis' shipmates had gone on to other pursuits around the Square and no one else was using his native language. The last Spanish he had heard was a few minutes ago when he and the rest of his shipmates had received orders over the plaza loudspeaker to report back to the boat post haste. A bother but he still had time to finish his cappuccino. Luis was a bit embarrassed about being called a 'submariner' because he was still very much in training and had not yet received his wings.

Luis said, "Si", and turned half away from his unwanted companion and tried to concentrate on his hot drink.

The stranger reached across the small table and clapped him lightly on the shoulder as though they were old buddies. "I know that look when I see it and I'm never wrong. You must be on the Hernando DeSoto tied down at the dock."

"Yes." Luis admitted grudgingly.

The stranger put a small package wrapped in a brown paper bag on the table, ignored it, and snapped his fingers at a wandering waiter. "A cognac for me and the submariner here, por favor."

He lit a cigarette and blew smoke up past his generous nose where, in the quiet air of the morning, it collected up inside the umbrella. Shortly, the waiter placed the drinks in front of them and Luis watched out of the corner of his eye as the swarthy one paid for them with American dollars. Luis was uncomfortable. The last time a stranger had tried to buy him a drink it turned out the man was inviting a homosexual encounter which sickened Luis. Not only was he lucky enough to be betrothed to the warm, motherly Estrella but he was a practicing Roman Catholic who really felt the tenets in his heart.

It was but a moment before one dread was replaced by another. The stranger pointed to Luis' drink and raised his own. "To you, Luis Martinez and your lovely Estrella, may you be favored with many sons." and he drained his cognac and sat there for long moments savoring the taste on his tongue and letting the impact of his personal knowledge of Luis sink in.

"How do you know my name?" Luis asked. He didn't really want to know but he had no choice but to ask the question.

The swarthy one turned his full gaze on Luis. His eyes were dark pits in his olive skin, like looking into the eyes of a snake. He smiled now but there was no warmth there.

"I know all about you dear Luis. You have been in the service of the King for five months, three weeks and four days." He looked at his watch then back to Luis. "Would you like me to tell you how many hours of the fifth day? Of course you don't. You have been a submariner apprentice now for two months and you are on that diesel boat down at the dock for a training cruise and your lovely Estrella is waiting for you in her parents apartment just off the Plaza de la Romano in Valencia. Her father will give you her hand in marriage when you truly become a submariner. Is it not true? Would you like me to tell you more?"

Luis felt something clinch him tightly at the pit of his stomach. He began to rise. "I must get back to my ship -- I am late."

The stranger reached across and put a very strong hand on his shoulder practically forcing Luis back onto his seat. He spoke lightly to Luis and smiled as though he were sharing an intimacy with a friend. "Luis, my friend, we have a favor to ask of you. It is in the best interests of you and your lovely Estrella to listen and to perform."

Luis forced the words out. "What has my betrothed got to do with this -- and with you?"

The stranger continued smiling but his voice turned to an ominous hiss. "The question is what we will have to do with your lovely Estrella if you do not do what we ask of you, Luis."

Luis blanched. "What do you want of me?"

The stranger pushed the small package toward Luis and smiled. "Open it."

Luis did with trembling hands and discovered a portable radio about half the size of a cigar box. He turned it over in his hands.

The stranger spoke softly, hypnotically. "You will take that aboard your boat and if anyone asks, you bought it as a souvenir here in Crete. The receipt is in the bag. When you return to that ungainly boat of yours you will put it under your pillow and forget it. When you begin hovering north of here this evening you will hear it come on. Ignore it and leave it alone until the batteries run out."

"We are not going anywhere but back to Spain," Luis said.

"Trust me," said the stranger. "You are going north."

"Is that all you want me to do?" asked Luis, much relieved.

"Yes," said the stranger, getting up.

Luis mustered his courage. "How will you know it is on?"

"How did I know so much about you, Luis. We have our ways. If you do that little favor for us, you and your lovely Estrella will live a long life and have the chance to have many sons. If not -- ah, but that is unthinkable."

"What happens if I turn the stations?"

The dark stranger smiled a warm, toothy smile, "There is only one." He stubbed his cigarette in the ashtray, said "Adios," and Luis watched him disappear into the crowd milling around the Fountain of the Lions in the middle of the Square.

Luis looked at the ominous little radio as though it were going to bite him. His hand shook more than a bit when he put the radio back into the bag. He didn't normally drink cognac but he ordered another and sat there nursing it for half an hour until the tremors left his stomach.

* * * * *

Number 1 was a very large man -- full of chest with strong sloping shoulders that made his suits not fit well – they were always too tight. He was too old, he reflected, for this kind of romp but he could not resist the glorious thrill in the power. He shrugged and shrugged and pulled and tugged at the suit until it reluctantly fitted itself to him and he could zip it over his paunch and up to his throat.

He checked the outer door again. It was securely locked. No one could see in, no one could get in so he would not be disturbed. He stepped through the inner door and gazed with pride at the shiny round cage embedded in the raised platform. It was sitting there like a giant Christmas tree ornament complete with the machinery that contained it, held it up and powered it. The small, intricate but very powerful SOTA computer that made it work sat in one corner. Massive wires ran through the wall to give it and the shell its power. He had brought it in himself, piece by piece. When he'd got it all together, a single workman, sworn to silence, had assembled it. At first he really didn't intend putting it to use -- he'd only cranked it up once. But he'd only left it on for a few moments because the lights

148

had dimmed and televisions had clicked off in the whole building. No, he just wanted the power of it. He wanted to be able to step into that inner sanctum and know that, if he chose, he could go any where in the world and have the power of life and death over any body he found there. How exhilarating it was -- the power of life and death. Better than the best orgasm he'd ever had. Made you feel like God -- hell, when you raised your right arm you were God.

He put on the helmet. It was uncomfortably snug across the ears. He'd ordered the biggest one they had and it was still too tight. He adjusted the chin straps as loosely as possible and tweaked the forehead panel to match the dials on the console next to the computer. He set the CPU on automatic, turned a knob on the console until he found the desired radio frequency then toggled a switch on the computer. The shiny sphere rotated two degrees and sat there waiting eagerly for its pilot. He walked up the steps and with some difficulty, slid his enormous bulk up through the door and into the rack then joined himself to the two umbilicals. Now he was part of the machine. He flicked a switch on his helmet and the hiss of the hundreds of jets rose until he felt total suspension within the womb of turgid air. He flicked another switch on his helmet and soared out of the building.

* * * * *

CHAPTER 26

"Has he come to yet?"

"No Commander."

"Doctor, did you examine him as I instructed?"

"Yes Commander, and I used smelling salts to no avail. Then I even poured some cold water up his nose but he did not stir." As an afterthought he added. "The lumps on his head are larger now than they were this morning. I think a concussion but without x-rays I cannot tell for sure. His left forearm is fractured. His blood pressure and heart rate are not good, but I am just a General Practitioner not a neurosurgeon."

"Damn! I told Paolo to knock him out -- not kill him!"

"He swears the American was about to squirm out from under the net and he had to stop him..."

"Shit! Paola just gets a kick out of bashing someone with that staff of his." He turned away from the Doctor signaling that the discussion was at an end. "Damn!" He snapped petulantly, "#1 is going to be calling me any minute to see what we've gotten from Chaney and I can't even talk to the sunofabitch."

* * * * *

Christa dropped her soiled clothes into the hamper and punched up the CPU on the intercom. The soft strains of Tchaikovsky's Andante Cantabile followed her as she made her way to the shower. She adjusted the stream to cool then stepped in and let the gentle water massage her.

As the tepid jets did their marvelous work she turned the water up as hot as she could bear. It soothed and caressed her body all the way to her feet. But the magic effect of the water only served to heighten her anxiety about Peter. She cupped her hands about her face and began to cry. Between deep choking sobs she leaned her forehead against the shower wall and prayed a simple prayer. Oh Lord, if not for me for Peter. He needs your help wherever he is, however he is, protect him and bring him back -- please.

* * * * *

Sorting through the mind of a true genius is akin to walking onto the beach at Coney Island at high noon on the Fourth of July and trying to identify everyone you found there. You'd never be able to finish the job because people constantly come and go. They're changing into different clothes, adding different makeup and getting sun burnt -- all changing their appearance.

A photographic memory and total recall is a gift on the upside. If you've read it, or seen it, or heard it you've got it and you rarely have to hunt for it. It's just there. It's something that everyone who doesn't have it wants it. But there are frequent down sides. Many times you're the only one in the crowd who has the answer and if it happens often, you have to exercise judgment or people end up hating you. Genius's frequently become reclusive because they can't find any conversation on their level. Can you imagine if the world spoke Russian and you were the only one who spoke English. Also as long as you're awake every door in your mind has something behind it and sometimes the clamor of all those thoughts overwhelms.

Even the genius himself sometimes can't cope. Jason's sixty year old body was exhausted, mentally and physically. He disdained a shower, instead plopping himself down on his bed and pulling the pillow over his head in an effort to shut out the maelstrom.

It didn't work. The soft, lilting strains of Tchaikovsky filtered through the pillow from the intercom. He recognized the work instantly It was The Andante Cantabile. But along with a photographic memory and total recall also comes a built in body clock so as he followed the melody he was not surprised when at one minute thirty six seconds into the score the music dropped to nothing and was replaced by a low pitched hum. That's the way the great one wrote it. But it persisted too long and seven seconds later when the melody should have begun to rebuild -- it did not come back. He rolled off the bed, snatched his cane and raced from the room.

As he neared the door to the CPU he heard crashes and booms and sounds like thunder and a male scream which was encouraging -- Paul was still alive.

Jason could smell the burnt odor of plastic and rubber and wood but so far no flesh. He peeped around the edge of the door into a lab full of chaos. Luci was standing in the middle of the room, laughing maniacally and throwing bolts at a cowering Ashton who had taken refuge in an alcove out of the direct line of fire. The corner of the wall in front of Ashton was on fire from Luci's repeated attempts to hit him. Smoke was pouring out of one console. Red lights were flashing everywhere. As Jason watched, Luci moved awkwardly across the room in order to gain a clearer shot at Paul. Luci had not seen Jason.

Jason screamed. "Cray!"

"Yes Mr. 'A'." Cray answered unperturbed.

"Give me the Star Spangled Banner at one hundred eighty decibels. Now!"

At the sound of Jason's voice, Luci turned and a malicious grin transfigured his face.

"Well Mr. Amador, I'm glad you've arrived. It is you I came to see."

Jason was already dropping as the monster raised that gigantic right hand and theatrically pointed his finger toward Jason. The bolt tore the door in half and blew it off its hinges. Jason lay looking up into the maw of death. Luci stood unopposed and lowered his massive hand to point at Jason down on the floor. Suddenly the thunderous strains of The Star Spangled Banner rent the room and Luci paused -- and the last Jason saw of the monster was that gaping, grinning rictus as it disappeared in the smoke.

Jason covered his ears and cowered there for a few moments trying to blot out the insane level of the music. When he was sure that Luci was gone he made his way over through the debris to the console and punched buttons to shut the sound down.

It took Ashton a moment to realize that Jason had defeated Luci then he grabbed a fire extinguisher and began spotting the flames. Jason turned around to see Christa battling fire on the other side of the room so he turned his attention to Cray.

"Cray."

"Yes, Mr. 'A'."

"Diagnose your damage and report to me please."

Immediately Cray came back. "One bank is down from a direct hit of seven thousand volts at twenty five amps..."

"Cray! Spare me the intensity -- just tell me what's wrong and how to fix it. At what capacity are you operating?"

"Eighty two point four percent. One auxiliary board is damaged beyond repair and must be replaced. You will need one AOX34 -- ABG board or thirty three million transistors, forty two capacitors, three miles of point eighteen micron copper wire and..."

"Enough Cray! Print me out a list of the parts I need to fix you."

"As you wish Mr. 'A'"

Jason turned his attention to a sheepish Paul Ashton who had just put the fire extinguisher back into its rack and was waiting expectantly.

He said quickly. "Jason, if you killed me I wouldn't be mad at you..."

Jason chuckled. "I won't kill you but I do want you to remember to take your music off tapes or CDS until this ordeal is over. I don't yet know how they determine who is on air and who is not but it *is* a neat trick." He paused for a minute. "Did you notice how awkward Luci was..."

Paul bobbed his head. "Yes, I couldn't believe it. He missed me three times. That's never happened before."

"Almost as though he was untrained, or a part timer." Christa said.

Jason turned around, surveying the damage. "I'm not going to mess with this tonight -- I'm going to bed. You youngsters can do what you want but in any event, Christa I want you to kill all the land lines coming in here. Malawi or Peter or Foley can reach us strictly through satellite".

* * * * *

El Capitan Espinoza sat on his bunk idly running a protractor across the Aegean Sea. He was thinking of episodes of Star Trek when the ship was in peril and Captain Kirk would call Scotty and demand 'more power, more speed'. Scotty would say with great apology "we're at one hundred ten percent now Captain, she'll blow if I give her any more."

Then Kirk would always come back, stalwart and strong and say, 'Scotty, it's now or never, give me one hundred twenty percent!'

He felt like that. Like the old Roman Captains in their slave driven galleys who only knew three speeds, cruise, flank and ram. Ram speed was everything the boat and its engine was capable of doing. It was flat out, a dead run, usually only sustainable for down the stretch runs. In a horse it brought the blood to a boil, the heart to near bursting and the breath to rasping jagged mouthfuls. Sometimes it got you to the finish line first -- sometimes it shook you apart before you got there. No matter you had to give it a try.

He'd told his Chief to damn the regulations -- give me ram speed

He could feel the hesitation in the Chief's voice, but a simple "Si" was his answer. So as he sat there in his bunk he felt and heard the frantic whump, whump of the diesels as they churned out speeds never before attained in this boat. In a boat that was designed to comfortably cruise underwater at nineteen knots and dash along at top speeds of twenty three he was flying one hundred fifty feet below the surface of the Aegean at ram speed exceeding twenty eight knots. As she did his bidding he felt as well as heard the old boat complain bitterly, creaking and grumbling in her seams. But if she held together this would place him on his designated point nearly two hours early. He had good use for that time. There was much to be done and there was the friendly civilian who was being held hostage to consider.

* * * * *

Jason paced around the room and finally stopped behind Christa's chair. She was busily coaching Cray. She stopped, looked at his reflection in her monitor then swiveled her chair. "Yes?"

Jason was unperturbed. He grinned. "In all your hacker wanderings have you ever been in the Naval Undersea Warfare Center Division in Newport, Rhode Island?"

She turned, pulled a small notebook from a drawer and flipped the pages. "Matter of Fact I have," she said smugly.

"Maaavelous," he said and took a chair facing her. "Have you ever heard of SOSUS?[44]"

She quirked an Eyebrow. "No. What is it?"

---

44 **SO**und **SU**rveillance System. A nearly worldwide system of hydrophones installed undersea by the U.S. Navy to track Russian Missile Submarines during the Cold War. Was highly classified information until after the cold war ended.

"During WWII German and Japanese subs raised hell all over the globe and all through the war they raised havoc principally because we couldn't track them nor find them. After the war was over and the cold war started it became absolutely necessary to know where Russian Missile Submarines were for very obvious reasons. In the mid fifties the U.S. Navy began laying an almost sea wide system of very sensitive hydrophones undersea so that we could track Russian Missile Submarines. It was classified until the Cold War broke, then declassified and now the Oceanographic people are getting Uncle Sam's money's worth out of the technology by using it to track things like thermals and whale calls."

Christa was getting impatient to get back to what she had been doing. "This is interesting, 'Mr. 'A' but what good is it to us?"

Jason grinned and ignored her light sarcasm. "Wouldn't it be something if you could go in there and find out where the closest hydro phones are to Piros and track that little sub running full blast toward her?"

He paused to let that sink in. Christa pursed her lips. "I can do it."

Jason got up off his chair. He said, "I thought you could. Let me know when you've got it sorted out."

* * * * *

CHAPTER 27

The tri-engined Sikorsky Sea Dragon is the largest helicopter outside of Russia. One   third of a football field long with its seven monster rotors cutting a seventy six foot swath through the air, it will cruise above one hundred seventy three miles per hour fully loaded with fifty five troops and their gear. With half a load its speed is classified.

Besides their size they are distinguished by a large canted tail fin which projects starboard and acts as an airfoil, increasing range and lift at altitude.

Like giant dragon flies resting on a lily pad four of them nested on the tarmac at the former U.S. Airbase at Hellinikon. Two of them had rotors turning lazily under test while personnel moved back and forth from trucks to the birds checking, testing, and outfitting everything from hydraulic fluid to ammunition. On each side of each fuselage enormous fuel holding sponsons shaped like life boats turned on their sides received umbilical cords from attendant trucks as fuel was poured aboard.   Three hundred yards away looking like baby dragon flies were six Cobras undergoing the same treatment.

Commander Joshua Templeton stretched his small muscular frame and continued to survey the busy scene from the third story of the control tower. There was some time to go before debarking and time was passing far too slowly for his liking. A brassy sun, starting on the second half of its day, painted lengthening shadows beyond the ships and men down below. It was a perfect day for a strike. He put a mobile phone against his ear and snapped, "Crew Chiefs report your readiness."

"#1 one hour forty minutes, Sir."

"#2 one hour forty five minutes, sir."

"#3 one hour forty minutes, Sir."

#4 one hour thirty seven minutes, Sir."

He couldn't help but grin. Master Sergeant Bukowski would always come in first regardless.

"XO[45], are you there?"

"Aye Sir."

"What's the status on the snakes?"

"Loading rockets now. They could take off in fifteen minutes, Sir."

"Excellent. All of you be prepared to load in two hours thirty minutes with take off in exactly three hours."

"Aye, Sir." came a chorus.

"XO?"

"Aye, Sir."

"Make sure the troops are in the Ready Room in fifteen minutes. I want to spend some time with them prior to take off."

"Aye Sir."

* * * * *

"Commander?"

He was sitting in his quarters in a comfortable seat in front of his desk idly watching a computer screen on which latest figures from the New York Stock Exchange were continuously being updated.

He toggled a switch on the consol next to his desk. "Yes Corporal."

"Please open channel 1, Sir. This video came in from Hellinikon just a few minutes ago."

"Efkareesto."[46]

He punched a button and the screen slowly came alive. The video was obviously shot with a camcorder and was hazy and off color. But the subject matter instantly made him forget the poor quality. Through the hazy heat he could see four American heavy helicopters being readied for toil and trouble. People scurried about like ants ministering to their four queens. Then the camera cut to a nearby Administration building from which a long line of battle ready troops snaked out across the hot tarmac. As he watched they split into four columns, one each to a different chopper and began loading. The camera cut to a wide shot showing a group of smaller Helios parked

---

45 Executive Officer

46 Thank You

in the background. He was wishing for a closer look when the cameraman went to his telephoto lens for a close-up and he could clearly make out six attack ships laden with rockets and Gatling guns. He knew this kind of armament was not routine and was only mounted under strike circumstances. I'd like to know their battle plan, he thought as the film ran out.

He toggled a switch, about to ask for a rerun, when the Corporal's voice cut into his thoughts. "Commander I have an eyewitness report that the four Cobra's just took off and made east northeast. They're loaded with Stingers..."

"They might as well swat us with tennis rackets.' The Commander snorted. "That would put them slightly north of us in about forty minutes. Corporal keep radar on maximum. Pick them up as soon as possible and maintain surveillance -- and keep me posted at five minute intervals."

"Aye, Sir."
"Any sign of the sub?"
"No Sir."
"Let me know when you do."
"Aye Sir."

* * * * *

Peter came to inside his head. A head full of hurt. The first rule of incarceration is that you are always under observation so he laid perfectly still with his eyes closed and listened. He was in a dark or dimly lit place that smelled of the sea. It was cool and the humidity was high. That placed him near the sea, possibly under it on the island. Memories came flooding back and he remembered the net and the painful fading blows. He could hear muted conversation and placed two voices. He was being guarded by two men. Beyond their voices was the hum of machinery and a distant bleep of a computer but not in this room. He lay perfectly still and willed his eyes to open a slit. His left eye refused the command -- it was swollen shut and hurting like hell. That side of his face was a mass of welts, lumps and scarified tissue having taken the brunt of the blows from Paolo's staff. The simple act of opening the right eye to a slit brought a renewed level of throbbing to the left side and he had to suppress the need to groan to relieve the pain. The first images his eye registered were double. He noted this with some anxiety, the first sign perhaps of possible brain damage. But as he allowed the eye to adjust to the weak light in the grotto his vision sharpened.

He was lying on a rough pad on a hard floor in the corner. The walls around him were black, unfinished rock which absorbed, rather than reflected, the single light hanging from the ceiling. Luck was with him. His face was turned to the right which gave him a view of the total room. His captors were in the opposite corner, about fifteen feet away, one sitting, one standing near the only visible door.

Sensing, more than seeing, a change in his prisoner, one of the guards walked over to Peter. As the guard began to move Peter closed his one good eye and steeled himself to stay loose. The guard looked down then drew back his foot and planted it sharply in Peter's ribs. The pain was excruciating sending red hot needles around his rib cage and deep into his belly. Peter called upon every ounce of his training in order to lay there and absorb the pointed boot and roll limply with the blow rather than stiffen and give away his consciousness. The guard watched him closely, then grunted and walked back to his cohort shaking his head.

Aided by the weak light, Peter began the long process of examining his body. He started by tensing his right foot enough to isometrically flex the muscles but not enough to make the foot move. He moved up to the right ankle, then to the calf and into the thigh muscles. Every muscle signaled back with its own brand of pain but the right leg worked. He progressed to the other side and found he had two working legs but a lot of pain in his left ankle. He imperceptibly pressed his left arm against his leg and drew his breath in sharply. There was sharp pain and movement between his elbow and his wrist -- movement that should not have been there. It was fractured. Damn!. He groaned inwardly. He was one armed.

He continued his assessment. He clenched the left fist through the pain in his forearm and found the fingers working but swollen and full of pain. The outside of the fingers on his left hand were skinned and swollen so he must have gotten them up to ward off some of the blows of the staff and that was when he got the broken arm. His left rib cage was a shambles. A deep breath brought its own painful reward and he noted probably two fractured ribs. Behind his ribs his back muscles complained and he dreaded the first flexing of the kidney area. If I get out of this I'm going to pee red for a week, he thought.

The left side of his head felt like a latex balloon had been inserted and blown up. The whole side of his face up into his hair line was numb and matted with blood. Try as he would he could not get his left eyelid to respond. The numbness extended down into the corner of his mouth, across the bridge of his nose and up onto the top of his head. He had no idea how the chip had been missed during such a beating. Since he was here, it was and he took that as a positive step in the right direction. If it was still there it was still working -- if he could find it.

* * * * *

The private line rang in the oval office. The President of the United States watched it ring twice then picked it up.

"Mr. President?"

He recognized the voice of Admiral Kincaid. "Yes Dennis, is this line scrambled and where are you?"

"Yes Sir, and I am aboard the USS Kearsarge just outside of Gibraltar."

"Are we in place and on time?"

"Aye sir. The birds are in the air on schedule and the explorer is about there and will be awaiting your bidding."

"Explorer?" It took a moment for the pun on the name of the submarine to register with the Chief Executive. "Oh yes -- just keep them standing by. With any luck we'll have closure on this in the next few hours and we can send them home. Any word on the hostage?"

"Regrettably no." said the Admiral.

The President sighed. "Keep me posted."

"Aye Sir."

* * * * *

CHAPTER 28

"Commander?"

He was in the toilet. He finished, flushed it and walked back into his office and toggled a switch on his console. "Yes."

"Hydro phones have picked up a diesel sub at our Southwest perimeter. Distance four thousand meters. She keeps pinging us.[47]"

"That would be the Hernando DeSoto. I've been expecting her, although..." he looked up at his wall clock... "she's a bit earlier than I thought she'd be. She's no threat -- we can take her out any time but just keep me posted. How about the birds?"

"They will be in radar range any moment now."

"Good. We'll wait and see what develops."

"Aye, Sir."

* * * * *

---

47 Sub Commanders sometimes will send a single Sonar ping at an object in order to establish absolute distance to the target. This can also establish the shape of the target since differences in the ping return time can chart the configuration of the target.

CHAPTER 29

Like busy bumblebees the four choppers churned a choppy path across the golden Aegean making for a point five nautical miles east and ten nautical miles north of the island of Piros from whence they would turn one hundred fifty seven degrees south, southeast for ten nautical miles, then come to two hundred seventy degrees west and bear in on the island. Not knowing how sophisticated electronically was his enemy, Commander Templeton had ordered his Cobras to wave hop so they flew in rough formation just twenty five feet off the water.

Anticipating radar from Piros he had also ordered the lead chopper to jam electronically and mechanically.

* * * * *

Jason sopped up the last dollop of his soft fried egg with his final morsel of wheat toast and addressed Christa who was loading the dishwasher. "Can't we patch through the nearest stationary weather satellite and get real time pictures of what happens when those choppers hit Piros?"

Without pausing Christa said. "None of the satellites are equipped with video -- but we can get stills."

"At what frequency?"

"They're solar powered so they probably don't shoot more often than one a minute -- maybe even one every five minutes."

"That won't help us." said Paul who was tidying up the sink.

"How about a consortium of satellites. Weather maps are pretty explicit. If you can get Cray to track and compile the shots from half a dozen satellites and present the pictures to us as one continuous array we aught to be able to see most of what transpires."

"I'll get right on it." Said Christa disappearing down the hall towards the lab.

* * * * *

"Commander, I have the four Cobra's slightly east and ten nautical miles north. They are aware they are being scanned. And I just have another eyewitness report of four Sikorskys and two Cobras taking off from Hellinikon . They appear to be headed this way. They're also loaded with Stingers.[48] " This from the senior of three operators manning the command console deep in the war room inside the guts of Piros.

"How do you know the northern group is aware of our radar, Corporal?"

"They are jamming, foil or something like it and they've just disappeared off screen."

The Commander surveyed the green radar screen over the shoulder of the senior operator. "Give me maximum zoom on the area where they just were and let the computer keep you on track on their most likely approach. Can you detect any outside radio communication?"

The operator punched a couple of buttons and pointed to a large screen on the wall above them. "Negative, both groups have been silent so far. The northern group were just in the upper left quadrant Commander. If you look carefully you can see a minute ripple in the sweep as it passes over the area. The other group is coming straight in to us..."

He pointed to the small screen where six distinct blips plowed slowly but inexorably toward Piros from the Greek mainland.

"Can you tell what they are and what's their ETA?"

The corporal consulted his small monitor. "I caught them as soon as they broke loose from the coast. I applied reverse tangents and that tells me this is the second group seen leaving Hellinikon. Four Sikorskys and two Cobra's. ETA with present course and speed about eighteen minutes."

The radar operator spoke. "I get a course change from the Northern group."

"What direction are they headed."

"South by southeast."

48"SAGW" Surface to Air Guided Weapon. Approximately six feet long and five inches in diameter. Can be shoulder fired or attached to Helicopters.

The Commander stepped back to get a better look at the wall screen. "That means they'll hit us shortly. ETA?"

"About fifteen minutes Sir."

He looked through the picture window in front of him into the huge hall where the three round cages gleamed dully. He could see his three cyber pilots each at rest in his cage. To the operator he asked, "Are the lasers ready?"

"Yes Sir.?"

The Commander grinned. It was the kind of grin that made you think he was the model for Luci. What a great chance to test the new technology and how fortunate we got it ready in time, he thought. Now they can't stop us from getting in if they get close enough to feel our lasers.

"Pilots, and operators -- are you at ready?" He asked into the Mike..

In unison came the answers. "Good" He said. "Each team will have at least three targets. This is the first test of our new technology and each of you pilots will enter on a laser which your operators will activate as soon as the target is acquired. I want you aboard, destroy and out as quickly as possible. No theatrics -- just kill the chopper."

He stopped and grinned. "Now there are ten targets up there so that means one of you is going to hit four and the team who does gets a hundred thousand drach[49] bonus each. Do you read me?"

Nine heads nodded eagerly as one. The Commander toggled another switch that opened all speakers to his Mike. "Listen to me. We will engage in approximately fifteen minutes. Their first group of choppers will attempt to hit us from the east and shortly after that the second group will come at us from the west. Pilots will destroy the choppers but I want you divers standing by your boats with full armament -- you will launch on my command. I want no survivors. All of you -- do your job and we will prevail as we have done in the past."

"Commander! I think there's a worm in the computer."

49The Greek Dollar. The Drachma. Depending on the exchange rate of the day this might be worth five hundred dollars.

The Commander turned to the Corporal, "What do you mean a worm, " he snarled.

"Something isn't right" the Corporal insisted. "I can't put my finger on it but every so often there is a glitch -- a blink of an eyelash that shouldn't be there. There is something happening in our system like maybe somebody strange is walking around inside it."

The Commander was attentive now. "Can you tell what it's doing?"

"No, I've run a systems check. Nothing is damaged, nothing seems to be missing. Everything is running as programmed -- I just detect a presence."

"How long now?"

"Maybe three minutes."

"Are backup systems online?"

"Yes sir."

"Switch to backup systems, NOW!"

The Corporal punched two buttons and flipped a switch. "We are now on backup systems."

"Any problems?"

The Corporal eyed his console, his eyes like saucers, "Yes Sir. The whole primary system just disappeared."

"We're alright aren't we?

The corporal consulted his board. "Yes Sir, all systems online."

The Commander smiled. "Could only be Amador -- is he in for a surprise. Switch passwords and fire walls[50] every thirty minutes. Now proceed as before."

* * * * *

---

50 An electronic wall designed to thwart illegal or illicit entry into your computer system.

CHAPTER 30

Inside the Sea Dragons the men spread out comfortably since only half the room was occupied. Although hastily formed as a cadre the SEALS had made an instant bond having all been spawned from the same training program and military ideals. Also having the chance to serve under a legendary commander had spurred many of them to answer the call. So as Templeton looked back over the charcoal scarred faces in his lead chopper he caught expressions mixed with anticipation, readiness for battle and awe at having been accepted for so important an assignment. In his pre flight remarks he had answered all questions forthrightly so every man could gauge the expected level of valor, the unforgiving price of failure and could sense the sweetness that would lie in success.

A calm voice on the intercom interrupted his thoughts. "Radar from Piros has locked on, Commander."

Templeton sighed. "Where are the snakes now?"

The pilot's disembodied voice came back, quiet against the chattering of the blades over head. "They should be cutting South, southeast now with an ETA four minutes before us."

"Does your radar pick up anything coming our way from Piros?"

"Negative. Nothing but the radar."

"Go ahead and jam. It won't help us now but may confuse them a bit."

"Aye sir."

* * * * *

Jason rotated his chair and faced Christa. "Is Cray having any luck breaking their code?"

"No -- but that's only half the problem." She said with finality.

Jason raised an eyebrow and Paul turned to listen. "Somebody wrote some smart code and the program is now self destructing."

"Damn. How are they doing it?" Jason asked.

"The binary structure is simply changing to all zeros."

"Useless. And how long before this is accomplished."

"Cray tells me we'll have fourteen million lines of garbage in just over an hour and there's no way to stop it without the key."

Jason shrugged. "Delete it Christa. Get rid of it before a monster pops out of it and tries to invade *our* system."

He turned and looked up at his wall screen. Cray was compiling information from every satellite that passed over the Aegean Sea, as many as twelve at a time, and producing a real time movement on the screen. She was concentrating over the Eastern Aegean and about one hundred miles of the western edge of Turkey. He set his cursor just west of the Island of Piros and double clicked it. Immediately the area came alive in great detail. Jason asked Christa back over his shoulder. "How far are we off real time?"

"Twenty second lag."

"Incredible."

He turned his attention back to the screen. Six black moving dots were approximately five kilometers west of Piros. To the east of Piros four black moving dots ceased their east by southeast flow and began to move abruptly west and were closing fast. The choppers were closing in on Piros. He hunched down in his ringside seat anticipating the clash.

* * * * *

The four Cobras hugged the bosom of the Aegean buzzing like angry hornets straight for the rock that was Piros. Three kilometers shy they fanned out into a flying wedge and prepared to engage at one hundred and fifty miles per hour.

Senior Captain Charles Bullway, USMC, with one bird to his left and two to his right broke radio silence for three words. "Fire at will" and triggered his first Stinger.

On the order three more Stingers disengaged themselves from the chopper's pods and streaked off for the target. Bullway, a hundred yards in the lead, and about to overrun the island pulled up sharply into the setting sun. As he lifted the chopper up through the smoke and over, his gunner behind him screamed, "My Gawd Captain, they've blown two of our birds to bits -- how're they doin' it?"

Bullway brought his ship to level just in time to see his third bird blow apart in the sky. Something awesome was targeting his troops. He rammed his Cobra up and over on its back ignoring all safety specifications then dropped it two hundred feet. Swooping out of the snarling dive, he loosed another Stinger at the island and hung there gimballing, raking the island laterally with his fifty caliber Gatling gun.

An intense light from the island struck him in the eye then probed his moving ship. A laser, he thought just as Luci materialized with a satanic grin inside the cramped cockpit. Then the chopper blew up.

* * * * *

"My Gawd, they're all gone!"

"What do you mean Chief?"

Admiral Kincaid quickly crossed the bridge to look over the Chief Bosun Mate's shoulder. The chief was looking at photos from the eastern Aegean Sea that had just come in off the satellite.

The Chief pressed the 'play' button then pointed to the screen. "Look there Admiral. We upped the shot frequency to one every second and you can see it happen. They all go out there and just blow up!"

The Admiral watched the screen in fascination. At maximum zoom, and following the Chief's pointer, he saw the four dark spots disappear in less than ten seconds.

He turned abruptly to the Captain of the Kearsarge. "Captain! Did we get any kind of radio transmission from those birds?"

"Negative, Admiral."

We're flying blind in this, he thought. The enemy has a super weapon. If they can do that to the Cobra's they'll do the same thing to the Dragons. Those boys are flying out there under orders to keep their radios off so Luci can't get to them which means we can't get in touch with them. He turned back to the Chief Bosun.

"Raise Hellinikon immediately. Scrambled line. Question: What's the Dragons ETA Piros?"

The Bosun twiddled dials, spoke briefly and turned to Kincaid. "One minute thirty seconds Sir."

"God dammit to hell!" snorted the Admiral and walked out onto the flying bridge and faced into the wind.

* * * * *

Peter heard the ecstatic yelling coming form the control room over the loudspeaker and willed himself to listen. He had periods of clear and periods of fuzz and sometimes he wasn't sure which plane he was in but he seemed to be slowly getting better. The mounting pain in his left arm was helping him focus. His two guards had been changed but were in the same position. His dungeon appeared to be down the hall a few feet from the main control room. Usually he couldn't hear anything but a slight buzz from there but this sound was different -- now it was an ecstatic roar. It reminded him of the sound in NASA Control Center when a launch went well. And then they cut the loudspeaker on so everyone on the island could get the good news. Within the din he understood that the first wave of Cobras had been decimated, destroyed unceremoniously, blasted from the sky and none of the Piros crew could stay in their seats such was the elation. There were no survivors. And apparently there was a second wave about to hit which would suffer the same fate.

"Sir..." the request was tentative amid the glorious uproar. Unseen to Peter the Commander waved his hand for silence but forgot about the open Mike... "What do you want Corporal , I've got more work to do!"

The Corporal put his hand to his earphones. "Pavel is requesting to speak with you direct."

"What does the Senior Programmer want with me at a time like this? How much time do we have before that second wave hits us?"

"Less then two minutes," came the terse reply.

The Commander looked around the room. "Is everybody clear on what to do"

A chorus of nodded heads answered him. He turned back to the Corporal. "Put him on."

The Corporal toggled a switch and motioned to the Commander who said with great irritation, "Well, Pavel. You've got less than a minute?"

The Senior Programmer was unperturbed. He spoke very succinctly as though he were using a knife to precisely dice a stalk of celery. He went straight to his point. "I am now able to beam an electronic frequency from point 'A' to point 'B'..."

"Dammit Pavel!" The Commander snapped. "Give me this in street language -- and quickly!"

169

On the other end Pavel smiled. He loved pricking the Commander's thin skin. It was always such a chore trying to explain new theory to a provincial. He translated in his mind then reiterated very patiently, "I can put Luci anyplace there is a receiving antennae..."

There was dead silence. The Commander stood transfixed trying to sort out this prophetic information. He drew a deep breath. "You're telling me that we need neither light nor sound to place Luci as long as there is a receiving antennae?"

"Correct."

"When will this be operational?"

"I can upload on your command -- you can use it in five minutes."

The Commander was ecstatic. He danced around like a five year old with two ice cream cones. "Jesus! What a breakthrough. Now nothing can stop us. We can blow missiles out of the sky -- we could catch a speeding bullet if it had an antenna. " He laughed out loud. "Do it," he yelled. "Do it! Do it!" He stopped and thought for a moment. "And upload to Number 1. Now that smart ass Amador can't hide from him."

Peter drew a deep, ragged breath. All seemed lost.

* * * * *

There was no sound in the lab except the omnipresent hum of Cray and her supporting machinery. Jason sat stunned. He could not reconcile what his eyes had just relayed to him. He switched into instant replay and watched the blips disappear again. Paul and Christa looked at each other in disbelief. Then with instant realization all three snapped back to the main screen. The six remaining black dots were about to breach the island from the west.

* * * * *

Captain Espinoza stood in the middle of the dive room trying to bring himself to terms with the information the radio operator had just handled him. All six Cobras were blown out of the sky. The first wave had failed miserably and every man was dead. The second wave was about to hit and would almost certainly suffer the same fate. He shut his eyes, said a quick prayer for their souls and thanked his patron saints. This was his hour of glory and he was here and ready. He gave two orders. "Get me Lieutenant Hopkins. Standby all torpedo tubes."

Hopkins appeared at Espinoza's elbow as if by magic. "You heard about your comrades?"

Hopkins nodded grimly.

"Piros will be after me very shortly. How soon can you hit the water?"

"Less than three minutes."

Espinoza turned a map around on the work bench and pointed to the south end of the island. He spoke quickly. "There appears to be a cavern in the south end of the island about fifty meters under the surface -- right there. I think it is a sub pen. Probably their way of retreat. We are the only ones aware of it. We are less than two kilometers from there now and fifty meters under the surface. As soon as you and your men are out of the escape hatch I will release your DPVs and orbit the north end of the island where I will attack with my torpedoes as a diversion. Your people need you topside. Take your men and God Speed."

* * * * *

Commander Joshua Templeton could just see the rim of the island twelve miles ahead as the choppers beat a path across the Aegean. His heart was pumping. In just under four minutes the high point of the whole trip would enfold him and he was ready. He turned and looked at his men. They were the best fighting men the world had ever mustered wearing the best equipment known in mankind's history and trained and imbued with the most time tested theories of attack -- and he felt secure. He felt that this was the great moment to which life had been leading him. The ultimate effort toward the ultimate quest for the good of his people.

At five kilometers out, he tapped the pilot on the shoulder and gave him the signal to 'fan' the choppers into the prearranged battle formation. The Dragons reduced their speed while the two Cobras raced ahead. The Cobras were to lead and continue the attack of the Cobras from the other side. They would attack also with their stinger rockets and drop smoke bombs on the surface of the island. They were to float, hover and sting, spraying any movement on the island with their fifty caliber Gatling guns. As soon as the surface of the island was deemed temporarily secure the Dragons would hover and disgorge their troops. The troops, led by him, would secure the top and work their way into the labyrinth taking prisoners if possible but bent on ending all resistance.

He braced himself, spraddle legged behind the two pilots, searching through the front bubble with his binoculars. He was looking for some signs of the four attacking Cobras on the other side. They should still be in engagement. He saw nothing but a faint haze of grey smoke drifting lazily downwind through the robin's egg blue sky.

Then suddenly, two hundred yards directly in front of him, just in front of the island, his lead Cobra vanished. It disappeared in a vicious, sight searing fireball and a huge pall of black smoke into which Templeton and his Sea Dragon plowed. With this his beautiful plan began to unravel.

* * * * *

CHAPTER 31

The six DPVs oozed out of the torpedo tubes like larva from a large cocoon. Charged to neutral buoyancy they hung in the hazy blue water as the SEALS flowed from the hatch and caught them. The DPVs looked like small torpedoes but with a small rear seat and flowing straps. It was but a matter of a few moments for one SEAL to board the seat and strap in, and the other three to catch a strap until six, four man squads churned off like giant cuttlefish, making for the island.

* * * * *

Templeton's Sea Dragon came out of the smoke almost on top of the island. All he saw around him were palls of black smoke and debris flying everywhere. He had a glimpse to the side as one Dragon bounced off the walls of the island and hit the sea in a flaming pyre. Other piles of wreckage were spotting the sea -- burning. He saw no other ships in the sky, he saw no survivors on the water. He turned to look out the port side and said "My God."

Luci was standing there, right next to him, inside the cabin, grinning maliciously. As Templeton stood transfixed, Luci raised his gigantic finger and whirled in a playful circle spraying crashing bolts of death as he went.

* * * * *

The Admiral, The Chief, The Captain and everyone on the bridge saw it at the same time. All six blips disappeared off the satellite photos. There was stunned silence. Everybody deferred to the Admiral and he said nothing. He was gripping the railing and his knuckles were white. He took a deep breath then said.

"Raise the Hernando Desoto."

In a few moments of silence the Chief Radio Operator said. "Admiral, I have Lieutenant Commander Espinoza, Captain of the Hernando DeSoto on the horn."

173

The Admiral picked up a Mike... "Captain Espinoza, This is Admiral Kincaid." His voice was very strained as though he was trying to lift two hundred pounds while he was speaking.

Espinoza's excited but calm voice came quite clearly through the speakers. "What is your bidding Admiral?"

"You are it now Capitan, Plan 'B' is now in full effect. Offload my SEALS ASAP and let them do their job."

Espinoza's voice was matter of fact. "They have been off my ship for almost five minutes now Admiral..."

"What?"

Espinoza's voice spanned the miles very quietly. "...Piros has known about me for at least two hours -- why they didn't strike sooner I do not know. But it is imminent any moment now. I felt the longer your men stayed aboard the less chance I would have getting them off intact. They are attacking through a cavern on the south -- WHAT IS THAT?..." He screamed! "...MADRE DE DIOS! EL DIABLO AQUI!![51]..."

The speakers went to hum with no connection on the other end. The radio man shook his head. "We've lost the signal Admiral, the other end just went dead."

"Get me the President, Chief. God dammit!" He stormed back and forth across the bridge. " I have to tell him we've lost ten choppers, one hundred and twelve men and now we've just lost a submarine full of children. It's been a day to eat rocks.

* * * * *

The Commander turned to his Number One operator. "What's the latest word on that sub?"

"Apparently out of commission Commander. "

"Can you get back aboard her?"

"No Sir. There's evidently water in that compartment because the radio quit operating.

"How about our new system?"

"Tried it," said the operator. "It won't function under water."

"Where's the sub now?"

---

51 Mother of God! The Devil is here!

"She's come to rest, dead in the water at about two hundred fifty meters. Must be on a shelf or something."

"What's crush depth on that old Pelosi Sub?"

The operator consulted a chart. "There's no recorded penetration deeper than one hundred eighty five meters."

The Commander smiled. "Well, if she's not crushed she'll sit there till what little air she has runs out then those SEALS and that kiddy crew will all go to sleep. Too bad,-- get me Number One. I have to tell him about our glorious day."

In a moment he had Number One on the radio and kept him spellbound for five minutes telling him about the one sided victory. He finished with, "You can't imagine how well it's gone -- by the way did you get Pavel's upload to you on the new program.?"

"Yes and I've got a couple of questions before trying to use it."

The Commander frowned. "Maybe you should talk direct to Pavel since we haven't used it yet either."

"Good idea. Incidentally, Set your fourth screen on to my equipment. In about thirty minutes I'm going to invade 'L'Aerie and get that bastard Amador and you can watch..."

"That'll put the cap on my day," chortled the Commander. "I'll love to see the look on Amador's face when his big decibels don't work. You say in half an hour?"

"Yes."

"Commander...?"

He turned, irritated at the interruption, about to read his subordinates the riot act but paused. Two of his wet suited divers stood before him dragging between them a bloody, dazed and bedraggled, more dead than alive, full Navy Commander.

"Please hold Number One." He turned back to the divers. " I told you no survivors," he snapped.

"Take a look at this one Sir, I think he's their leader."

The Commander stood up, looked closer but not close enough to get any blood on his own uniform. The prisoner was small in stature but wore the silver oak leaf of a full Navy Commander. Blood covered the side of his head, was spattered all over his uniform and water and blood dripped into a puddle off each limp leg.

"We thought you might want a hostage Sir. We found him down among the rocks where that one chopper crashed."

"Throw him in with the other one." The Commander instructed, "I've got other more pressing problems. I'll kill them both later."

He went back to Number One. "D'jou hear Number One?. Now I've got two hostages."

"Has Chaney ever come around yet?"

"No."

"No matter, this is all but over now anyway. All we have to do is tie up loose ends and disappear and it's over -- watch the old man in thirty minutes."

"Will do."

* * * * *

Jason, Christa and Paul were clustered around Christa's monitor and also watching the wall screen. She projected a picture of the floor of the Aegean onto the big wall screen. She pointed her cursor. "Look, there's one phone on the edge of the Anatolian Trough here in the northeast, eighty miles from Piros; there's another one down here between the coast of Greece and the island of Skiros about fifty miles from Piros; and there's a third one over here on the west coast of Turkey less than twenty miles from the island. I've got Cray programmed to lose all normal sounds like whales, dolphins, fish and surface vessels -- although I don't think there are any of those around there right now." She paused and turned to Jason. "How fast does sound travel in sea water?"

"Meters, yards or feet?"

"Dammit Jason, just give it to me."

Jason chuckled. "A little less than a mile per second. Five thousand and twenty three feet."

"The feed from the Turkish phone is going to be the closest to real time and that means it will be at least twenty one minutes behind. Cray is registering those sounds now..."

Jason said, "Cray, identify those sounds please."

Cray responded instantly. "First ten seconds is that of a medium sized, single screw, diesel submarine, approximately fifteen hundred tons, speed approximately ten knots. Next forty three seconds indicates severe internal detonations, forward motility ceased, engines shutting down and water invading the hull..."

"She's dead in the water and sinking." Jason cut in.

"That is correct," said Cray.

"Dammit," grimaced Jason. "There's twenty six youngsters, five officers and twenty SEALS aboard that ship and she's sinking in fifteen hundred feet of water. That's almost three times her crush depth. Cray,"

"Yes, Mr. 'A'."

"Keep filtering those sounds. When that sub implodes it should make a distinct sound. At least we'll know what happened."

"Yes, Mr. 'A'.-- But I am registering sounds from the sub that are definitely not those of implosion..."

"What are you hearing," Jason snapped.

"At approximately two hundred meters below the surface I hear the sound of metal impacting stone."

Jason thought for a moment. "Cray, does that sound continue in depth or does it stay at that level?"

"There is no evidence of sound below two hundred meters."

Jason looked at Paul and Christa. "Maybe the day is not yet lost. Sounds like that sub has lodged, at least temporarily, on a shelf at two hundred meters..."

"That's below crush depth, isn't it?" Paul interjected.

"Yes. But those are engineering estimates anyway. With good rivets, good steel and good workmanship it might be capable of going to twice the crush depth without imploding. However, we can't stake their lives on that. Cray.."

"Yes, Mr. 'A'"

"Keep listening for any sound from that sub and notify me immediately if you hear anything."

"Yes, Mr. 'A'. Wait, I am getting more sound."

Thirty anxious seconds went by as the three waited for Cray to speak. Finally she said, "I am getting metal tapping against a metal hull. It is International Morse Code -- in Spanish. It says simply "Nosotros vivir."

"My God," Jason exclaimed. "We live!"

* * * * *

The guards dumped a limp Templeton unceremoniously next to Peter. The guard nearest Peter gave him a half hearted kick in the ribs but was so used to no reaction   that he did not pay close attention to

Peter. Both wandered back to the door and outside where they could join in the festivities.

Peter got a grip on the pain then opened his one good eye to just a slit. Two feet away from him he was staring into two blue eyes which were startlingly clear considering they were set deep under a mat of bloody hair and a face covered in soot mixed with blood. He was a small, compact man, not nearly so tall as Peter.

"Chaney?" Templeton spoke with difficulty but said it so quietly that he almost just mouthed the words.

It was the first time Peter had tried to speak since being captured. It was not easy to make the right side of his mouth work and his throat felt thatched with dry hay. He swallowed and rasped. "Yes, who're you?"

"Templeton, Navy SEAL. I crashed up above -- all my men gone. You and I are it."

He looked Peter up and down as much as he could without being observed by the two guards just outside the door. He noted Peter's face. "Lotsa damage, you still intact?"

Peter smiled with the right side of his mouth. He liked this little man -- he went straight to the point.

"Left arm broken -- you?"

Templeton grimaced in sympathy. "Cuts and bruises -- I can walk."

Peter said. "I have secret weapon..."

Templeton looked at the horrible scabs on Peter's black and blue face and chuckled quietly. "Yeah, you sure look like you do."

"I have, " Peter protested slowly over his swollen tongue, "I tap my head three times and throw lightening bolts."

Templeton looked at Peter and sighed. He's really out of it. Man -- I am all by myself.

"Well, if you can you'd better crank them up, 'cause they're gonna kill your old buddy in about another twenty five minutes."

"Wha...? Peter was having trouble concentrating.

"Amador's your buddy isn't he?"

"Yes -- s."

"Well I overheard that bastard out there in the control room talking about how they were going to invade his place and -- they've got a new system. They're going to kill him and sound won't stop them this time..."

"Jesus...!"

* * * * *

They could see the lightness ahead through the pristine water long before they reached the mouth of the cavern. Hopkins waved 'dead slow', then stop, detached himself from his DPV and swam cautiously forward. The island thickened as it dropped toward the ocean floor. He judged the entrance to be thirty meters wide and at least forty meters high and it was totally under water. There was a submarine net stretched across the opening from wall to wall but only about twenty feet high. It was sufficient to stop another boat from entering but no barrier to a diver. He stopped and spent precious moments looking for sensors. Finding none he motioned his men to follow and swam cautiously inside.

As he entered he motioned his men to stay below the surface. He hugged the inside of the front wall and levered himself slowly to the surface. He immediately felt the air pressure that kept the sub pen only partially filled with water. He turned his head and gasped. The audacity of what he saw astounded him. He was looking at the stern of a diesel submarine approximately two hundred and fifty feet long and probably displacing at least twenty five hundred tons. It was shoehorned into a convex roofed cavern that was just long enough, wide enough and tall enough to accept the boat and have room on each side and the front end for absorbent bumpers. The inside of the cavern had been lathed with mortar, making it airtight and grey instead of the dark, natural rock color. The roof stretched above the conning tower perhaps fifteen feet. Two feet above the sub deck there was a steel catwalk, five feet wide, attached to the wall of the cavern and it extended around all three walls. On the left wall a waterproof door opened onto the catwalk and a short gangway angled down from there and rested on the deck of the sub. He could see both fore and aft hatches were open and assumed the conning tower hatch was as well. He could hear the muted cacophony of Rock 'n Roll. There were no guards visible.

He dropped back below the surface and gave detailed hand signals. He pointed three men to monitor the hatches and the conning tower and pointed to three others who carried magnetic, thermite, shaped charges[52]. He made unmistakable motions of attaching them to the hull.

The three men detached with the explosives and went straight for the sub. They moved silently up under the hull. One paused just in front of the propeller shaft, the second swam forward under the full belly and the third went three quarters of the way toward the nose. Each silently laid his charge up against the hull and pulled a pin. The charges were set for ten minutes. The other three, looking like dark suited Ninjas in their wet suits and rubber soled shoes, sped up the side of the sub. One went for the first hatch, the second scurried up the conning tower and the third made his way forward to the forward hatch.

Hopkins led his men to the port side of the sub and gained the bottom of the gangway. He looked up at the conning tower and received the 'all clear' from the SEAL up there. He ran swiftly up the gangway and inspected the massive door. It was built to withstand tons of water and thousands of pounds of air pressure and it would have been a loud, long and formidable task to blow it. He rotated the latch wheel and found it unlocked. There was a whisper and a grunt from the door as he eased it open, knife at the ready. Inside was a short, empty hallway with just enough room to cram fifteen men. There was another airlock at the other end. On the right wall was a panel of switches and lights. He knew that both doors would not open at the same time -- they were the only barrier to the tons of water between here and the surface. The question was, would that other door open once this one was shut or would they have to blow it. He held up his clenched fist[53] stepped in and let the door close behind him. He rotated the wheel fully closed then stepped quickly to the other side and grabbed the wheel. It turned easily in his hand and the door opened with a corresponding wheeze and a grunt. He looked out onto a long, dimly lit, uplifting, empty hallway. There was only one

---

52 A 'shaped charge' is designed to explode in one direction only. In this case it is designed to drill a hole straight up through the sub's belly and spray white hot thermite all over the inside. A violent fire followed by explosion will be the result.

53 The military signal to 'hold -- assemble'

way to go. He turned back, closed the door and noticed the flashing red light on the panel as he passed it. Time was running out. He assumed that signal was also flashing somewhere else on this island and they would know the doors had been opened.

He stepped back through the first door to see relieved looks on the faces of his men. He grinned and waved at the three SEALS on the sub. The three scrambled over to the gangway and were joined by the three who had come from underwater. They followed the main group into the airlock.

The SEALS crammed like sardines into that hallway. When the end man finished screwing the door shut, he gave the signal and Hopkins, in the lead, who was jammed up belly tight against the wheel, began turning it. It turned harder this time with his weight against it but turn it did and the door wheezed open.

He pushed his bulk up the hallway with his men streaming eagerly behind. The last man did not stop to lock the door down and the red light on the panel went crazy.

* * * * *

CHAPTER 32

"Commander!" The Corporal said. "Have you ordered any of the sub crew topside?"

"No -- why?"

"Both doors were opened and then closed."

"You sure it wasn't a malfunction?"

"It's never happened before," The corporal said defensively.

The Commander stepped to the door leading into the main cavern. A dozen divers, Peter's two guards and other assorted ruffians were standing around in a gaggle enjoying their recounts of the victories. They stiffened as the Commander yelled. His wave encompassed them. "All of you, on the double with weapons. Get down to the sub pen and see why those doors were opened -- move!"

He watched them disappear at the end of the room through the doorway which led to the lower passages. He turned and grumbled his way back into the control room.

As he reached his seat the Corporal looked past him and his eyes went wide with terror. The Commander turned in time to see a five foot six inch bloody apparition standing there who promptly hacked him across the nose. His ear phones and throat mike went flying and the ferocity of the blow dropped him to his knees. He looked up and the bloody little bastard in front of him, who was hardly taller than he was kneeling, was bringing his rigid hand down in a murderous slash which caught him full in the Adam's apple, crushed his larynx and stopped all his air. He fell over backward.

In the meantime, Peter made it to the door of his prison and was standing there unsteadily searching the top of his head. He was looking out on the main cavern where the three cyber pilots were busy applying Luci.

Dried blood, loose tissue and even some small pieces of seaweed made it difficult for him to find the SAM. Finally he found a spot that seemed right and he tapped it three times. His skull was so tender the light tapping brought tears to his eyes. He started counting -- one thousand, two thousand, three...suddenly he went rigid and he looked down at his right arm. He could see an aura surrounding it, almost like a cast made of clear water. He moved it slightly being careful not to point it yet...five thousand, six thousand... he waited, he wanted a full charge...ten thousand, eleven thousand, twelve thousand...he raised his arm and pointed it at the last cage...fourteen thousand... he raised his index finger.

There was a shattering response from his finger to the cage and the cage disintegrated in a cloud of smoke and debris. He reached up and tapped his head again.

Inside the control room Templeton was a deadly whirling dervish. He caught two of the operators with telling blows as they sat transfixed in their seats and he now sat astride the terror stricken corporal about to administer the coup d'etat.

There was a startling explosion out in the main cavern. He looked up as one of the cages disintegrated before his very eyes. He looked around for a source and all he could see was a bedraggled and very unsteady Peter Chaney tapping himself on the head. Templeton gripped the corporal's throat tight enough to stop the Corporal's activity and watched Peter in fascination. Peter stood absolutely still for a few seconds then stood bolt upright as though someone had suddenly tied him to a straight pole and an iridescent halo surrounded his whole body.

In the sudden silence he seemed to be counting. Templeton could distinctly hear him saying "...nine seconds, ten seconds, eleven seconds..."

He watched Peter extend his arm and then methodically point his finger. There was a loud zap and a bolt of purest blue jumped from his finger and sundered the second cage.

* * * * *

Sprinting silently on rubber soled shoes the SEALS wafted like deadly black ghosts up the stairs, around the corners and through the doors. They encountered but two guards and those were lackadaisical and dispatched with little effort. They heard the shouting of the main group long before they got close and were hunkered down in the passageway from whence they delivered fatal fire as the other group came pell mell around the corner. The SEALS poured through the door to the main cavern in time to see Peter loose his pyrotechnics on the second cage.

* * * * *

There was one muffled scream from the terrorist's sub. Had anyone been listening they would have known it was a man aboard but it sounded like mortal pain from the metal beast itself. The scream pierced the damp air for perhaps three seconds, then there were a series of whumps and white hot thermite flames shot out of the three hatches. Five seconds later the belly of the sub opened like a gutted fish and jammed the conning tower against the roof of the cavern. The resulting pressure wave drove half the water back out into the Aegean and blew both air locks off their hinges. An inferno of flame and acrid smoke surged through the airlock and up the passageway in pursuit of the departing SEALS. The cavern belched, then inhaled hundreds of tons of angry Aegean bent on reclaiming its rightful spot within the island. The water doused the flames and at two hundred miles an hour roared up through the open passages looking for sea level.

* * * * *

Templeton looked past Chaney and with great relief saw the SEALS burst through the far door. At the same time red lights began to flash in the Main Control room and a klaxon began its insistent bleat. Templeton looked around but continued to hold the Corporal by the throat and drug him over to the door. He yelled at the SEALS. "Hopkins, Commander Templeton, over here."

At that moment Peter, more out of his head than in, reached up to the top of his head and began to tap. Templeton made a short lunge and caught him before he hit SAM the third time. Peter gave him no argument but slid down to the floor and looked groggily at the SEALS who were now standing around waiting for orders. Templeton grabbed a Sergeant standing there and said, "Do not hurt him but do not let him touch his head, understand."

Looking very puzzled the Sergeant nodded.

The Corporal writhed under Templeton's grasp and frantically tried to speak. Templeton tightened his grip and drew back his hand to hit him. Sounding like a crow with the croup the Corporal rasped, "The watertight doors -- shut the water tight doors."

"What for?" Templeton demanded.

At that instant a gout of air burst through the far door. The corporal, dangling from Templeton's grasp like a minnow on a hook this time literally screamed, "We've been breached! The sea is coming in -- we'll be under thirty feet of water -- let me shut the doors." and he pointed to the console. Templeton shrugged -- the Corporal had no place to go. He pushed him at the console and watched him slam the switch marked "WTD".

Through the far door he could see fifty feet of straight passageway up which the SEALS had come not two minutes before. As the door began to slide closed he shuddered. A wall of water surged around the far bend and slammed against the other side of the door as it settled into place. It creaked and groaned and little sprays and sprinkles forced their way under the door. He prayed that single door could hold the hundreds of tons of water that so desperately wanted to reach equilibrium.

"Hopkins," he shouted. "Drag that cyber pilot out of that cage. If he fights you, kill him!"

He turned to the Corporal and slammed him to his knees. He raised his hand in a killer chop and said, "Do you want to stay alive?"

Terrified, the Corporal nodded his head.

Templeton smiled like a small ghoul, "Here's what you're going to do for me."

* * * * *

"No, Mr. President, we have heard nothing since then to indicate that they are gone or dead. The next question is how fast can you get help to them?"

The President was matter of fact. "Mr. Amador, hold on and let's get Admiral Kincaid in on this conversation..."

Jason kept the phone to his ear but shrugged at Christa. Soto voce he said to her, "It's almost lunchtime, how about rustling something to eat?" She grinned and left the lab.

In a moment the President came back on the line, "Amador, you still there?"

"Yes, sir."

"OK, Admiral Dennis Kincaid is our third party and this is a safe line. Tell him what you've just told me."

Jason did so and ended with the same question he had posed the President.

Kincaid said, "Two hundred meters is too deep for the Steinke Hood.[54] So we'll have to go with a DSRV[55]. Luckily I've got one nearby in the English Channel. that I can fly to Athens in six hours. It can go the short distance over to Piros by itself and start work. I've got a nuker, the Marlin, in off Gibraltar that I can ram over there in time to mother the DSRV. But these problems are not our basic problem, Gentlemen. How do we stop Luci from wrecking the DSRV?"

"Secrecy will allow them to get there Admiral. Luci obviously can't work under those conditions or they would have finished the sub off. The DSRV would have to maintain absolute radio silence and stay submerged all the way from Athens to Piros and back out of there. Besides we'll have six more hours to figure a way of combating Luci..."

The Admiral was envious. "It's amazing you've been able to come up with this information when the Navy has had access to that system for nearly fifty years. Has the sub given you any indication of the condition inside the sub as relates to oxygen concentrations, carbon dioxide concentrations, internal pressures, crew injuries...?"

"Nothing Admiral -- nothing but the International SOS and the two words 'We Live'. Obviously there's no way to let them know we hear them so we can't ask questions. "

The President asked, "How often are you getting messages from them?"

"Every fifteen minutes an SOS."

The President chimed in, "Dennis, any word from your SEALS on Piros?"

Kincaid answered wearily, "At this moment Mr. President I haven't the slightest idea what's happening over there. They don't have radios strong enough to reach us. For all we know they could all be dead."

---

54 The standard Individual Escape Device carried on most submarines. Safe to be used down to four hundred fifty feet or one hundred thirty eight meters.

55Deep Submergence Rescue Vehicle. Can be used as deep as five thousand feet.

"Too bad, that was our last great opportunity...," grumbled the President.

The Admiral signed off, "Yes Sir. I'm going to put that DSRV in the air. We'll sort the rest out later. The first thing we've got to do is try to rescue that Captain and those kids. I'll keep you posted Mr. President."

Jason also signed off and sat there reflecting on the days events. No word from or about Peter. No word from or about Piros since the last wave of choppers went in. No word from or about the twenty SEALS off loaded from the Hernando DeSoto. No solid word about the sub. Truly a day to burn in effigy.

He suddenly noticed the hair on his bare arm stood straight up, for a moment he could feel each hair on his head and a chill ran down his spine. He turned to ask Paul if he had changed the thermostat and froze.   Thirty feet across the room, grinning maliciously, malevolently, evilly stood Luci. His gross python tail curling up and around his gleaming torso, his muscular snake's neck undulating, his great Preying Mantis skull tilting this way and that. He had his great hands at his waist resting gently. He made no move to raise his right arm . He just stood there glaring at Jason -- enjoying the moment. Jason looked at him -- studied him. There was something different -- something changed about Luci. Then he saw it -- a small protrusion between the horns, like a miniature antennae. Jason had no time to reflect on this as Luci broke the silence. Paul, with his back to him, jumped a foot above his chair at the sound of his sepulcher voice.

"Your time has come Mr. Amador..." He turned toward Paul, "... and yours also Mr. Paul Ashton. The two of you have been a minor thorn in my side for days now -- but no more. Today is reckoning day and I am going to excise both of you and be rid of the problem..."

He did a tiny little pirouette as if to show that he had conquered his previous bungling and his grin got bigger. He said condescendingly, "Don't bore me with your decibels, Mr. Amador, I no longer depend on sound to get here so loud sounds don't bother me any longer..."

He suddenly raised his arm but did not point the dreaded finger at Jason -- he pointed it at Paul. There was a thunderous crack and this time he did not miss. A neat, little hole appeared in Paul's forehead and he fell over backward on his console and rolled to the floor.

Jason had moved as Luci's hand came up thinking the bolt was meant for him. He snatched his cane as the only weapon, however ineffectual, and rolled behind a desk.

He waited for what seemed hours for the bolt, but none came. Not a sound in the lab.

He snuck a peak around the desk and there was Luci, standing stock still, consternation written across his evil face. The only movement was his tail which whipped around him in frustration. He was staring past Jason at the opposite corner -- and not making a sound.

Jason pulled back behind the desk and turned his head and glanced at the opposite end of the lab. For a moment he thought Luci had found some new method of locomotion because there he stood grinning like a devil. But a quick look back confirmed the original fact -- there were two Lucies -- and he was between them.

He looked back at the second Luci who was also grinning -- but this grin was different -- there was mischief in it.

The second Luci broke the silence. "Give me your best shot Number 1. Try a bolt on me instead of this defenseless old man..."

Luci One raised his hand and both Lucies shot a bolt at the same time. A double clap of thunder that rent the room. Both bolts went straight through their respective targets and bored holes in the walls behind them. The walls around the holes began to burn and flames began to eat at the inner walls.

Through the smoke, Luci Number One grinned his evil grin. "I don't know how you've managed this but your bolts don't faze me -- I'm still going to kill Amador and you can't stop me..."

He shot another bolt at Jason but distracted by a bolt from Luci Two he missed. Luci Two fired two more in quick succession then closed with Luci One. He surged right through Luci One who laughed in glee and fired again at Jason. The smoke was now so thick it was difficult to see. Jason discovered Luci did not have supernatural eyesight and suffered in the smoke as well. He slid under a console and pulled a mike in with him.

"Cray, can you hear me?" he asked very quietly.

"I hear you Mr. 'A'. " Jason could also hear and see Lucies One and Two stumbling around in the smoke. When they surged through each other the air was doubly charged with electricity.

"Cray, If you charge every capacitor you've got to maximum how many volts could you give me?"

"Three hundred twenty five thousand, six hundred and forty two for five milliseconds."

"Do so and standby to discharge them through your spare com port."

"I will need fifteen seconds."

"Do it."

"That amount of electricity will fry whatever it strikes, Mr. 'A'." Cray said conversationally.

"I'm counting on it," Jason said grimly.

While he watched the second hand of his watch Jason unscrewed the leopard's head from his cane, ejected the two forty five shells from the chamber of the derringer, then screwed it back together. When twelve seconds had transpired, he stood up in front of the com port and yelled into the smoke. "Luci Two stand aside, this is between me and Luci One." In final preparation he stripped the end from the cane revealing the deadly fifteen inch blade.[56]

A thunderous laugh greeted this Quixotic stand and Luci One, drawn by Jason's voice, came flouncing out of the smoke. He was flourishing his right arm, raising it high above his head and playfully making as if to steady it with his left as he brought it down, much as you would a sniper's rifle.

Jason gripped the wooden shaft of his cane, sighted along it to Luci Twos head and jammed it backward into the com port.

He was not quite sure of the next few moments. He knew there was a monumental blast of electricity that breached the distance between the end of the rapier and Luci's head. It charged the air, held Jason in place, made his mouth feel like he'd brushed his teeth with marble dust and turned his cane into a red hot poker that he could not drop before it burnt his hand. Every hair on his body protested and his crowning glory on his head turned brittle and fell off in little shards. But with great satisfaction he watched Luci's head turn a mixture of

---

[56]Jason Amador always carries a solid gold Leopard's Head cane. The top part of it conceals a two shot forty -- five caliber derringer and the bottom half of it conceals a fifteen inch razor edged rapier.

bright, and white, and light blue as the staggering bolt of juice centered on the little antennae between his horns. .

Then Luci One disappeared as he had come.

The blast lasted only the blink of an eye. Then it let Jason loose and he fell back on the console then fell over on the floor. He had no feeling in his arm. All the hair was singed from his head and his chest felt like he'd been rubbed down with a cheese grater. His heart was racing and the blood was roaring in his ears. From the floor he saw movement out of the corner of his eye and was vaguely aware of Luci Two materializing through the smoke.

"Who are you?" He asked groggily through a very dry mouth.

Luci bowed low and brought his hand across his chest as though he held cap with feather. "Another time, another place, Mr. Amador, now I have no time..." and he was gone.

THE END

EPILOGUE

Christa manned the fire extinguishers and saved what was left of the lab. With many tears she called local law enforcement who took the remains of Paul Ashton and shipped them to his sister back in California. She cuddled Jason's bald head until the ambulance carted him off to the hospital. The next day she brought in selected workman and started L'Aerie back to its original condition. She also bought Jason a wig and presented it to him at the hospital.

Jason spent two days in the hospital recovering from shock and wore his arm in a sling for a month until the nerves refreshed and he could move it comfortably again. It took over a month for his fine head of hair to grow back and Christa was heard to chide him about growing bald. His hand pained him for months before the tingling subsided. He sent his cane to the London sword maker who had produced it but they convinced him he needed a new one rather than try to rebuild the burnt out core he sent them.

Peter Chaney was airlifted by chopper to Athens where he was met by a comely nurse and a beautiful Doctor (both selected by Jason) and flown back to the United States in Jason's jet. It would take three weeks in the hospital, one jaw operation to restructure his teeth on the left side of his mouth and another to put a pin in his left arm and then a month at home in his California apartment before he was ready as new.

Commander Templeton made it back from his impersonation as Luci Two just in time to move his group out of the Main Cavern and up to the top level before the airlock crushed in under the weight of countless hundreds of tons of water. The water rushed in and destroyed and buried all the data thirty feet deep. Templeton's valor was noted by the President and in much publicized ceremonies Commander Templeton was promoted to Rear Admiral Templeton. America had a genuine a hero.

The DSRV was flown into Athens and then out to the wreck site but Luci had done a thorough job on both the Hernando DeSoto and her crew. Only five men were brought out of the crippled ship. Luis

Martinez was not among them. Sadly, neither was El Capitan Santiago Arredono Espinoza. The King of Spain received a personal letter from The President of the United States outlining the length and breadth of Captain Espinoza's contribution to world peace and was assured that his bravery and dedication would always be considered whenever Spain and the United States had occasion to discuss world conditions. It was decided by both parties that the sub should remain the final resting place of the Capitan and his men so a month later, as a courtesy, the United States dropped a diver onto the wreck, placed appropriate explosives and nudged it off into the abyss.

Three days after the day all the lights dimmed and the TVs fuses blew in the prison building (for the third and last time) Jed Foley, "The First Associate", still hadn't shown up for work so the Assistant Warden broke into Foley's outer office -- and his inner office -- and found an indescribable apparition hanging in a funny contraption that looked like a giant Christmas tree ornament. The figure was in the shape of a man but had been fried to the point where all the oil and water had been rendered from the body and it had been mummified like Ramses II. DNA results confirmed it was Jed Foley. All the hardware connected with Mr. Foley, aka "The First Associate, aka 'Number One', was charred and brittle and all the software had been consumed in the flash so there was no record of his activities. The prison electrician opined that Foley's office had been hit by a bolt of lightening although weather reports for the day showed a clear sky. They couldn't tell Mrs. Foley as she had disappeared before they started looking for her. Digging further in her records it seems her family name in Greece had been Metaxis and her father had changed it to Martin to Americanize it when he had brought his family to the United States.

As Jason said it would be, the money was never traced. And it was all about the money. The bacteria, and Luci both were elaborate, murderous ploys designed to get the money for the family.

REALLY THE END

Bibliography:

"A Field Guide To Airplanes," 2nd Ed. M.R. Montgomery/Gerald Foster, Pub Houghton, Mifflin Company.

"Bracey's Land Warfare, Guided Weapons," 3rd Ed., R.G. Lee, T.K. Garland -- Collins, D.E. Johnson, E. Archer, C. Sparkes, G. M. Moss, A. W. Mowat.,Pub Brassey's London.

"Submarine," Tom Clancy, Pub Berkley Books, New York.

"Learning To Fly Helicopters," R. Randall Padfield, Pub Tab Books, Div of McGraw --Hill.

"Modern U. S. Navy Submarines," Robert and Robin Genat, Pub Motorbooks Intl.

"Modern Military Techniques, Submarines," Tony Gibbons, Pub Lerner Publications Co, Minneapolis.

"The Spanish Vest Pocket Dictionary," Edited by Donald F. Sola, Cornell University, Pub Random House, New York.

"Greek For Travelers," Berlitz

"FAS Intelligence Resource Program," U. S. Navy, Internet.

"The Mediterranean Seafloor" Map produced by The National Geographic Society.

"Naval Ships Tech Manual, Chap 594. Salvage -- Submarine Safety, Escape and Rescue Devices, Published by Direction Commander, Naval Sea Systems Command. U. S. Navy.

ABOUT THE AUTHOR

The Male 'Grandma Moses' of mystery fiction, Rus Morgan has been a marine, an account executive, a laborer, a poultry and cattle farmer and owned a contruction firm. He is a Mensan. He has published Investigative Feature Articles in the Memphis Magazine, The Flyer, The Memphis Business Journal, 'Hand's On Electronics' and The Hollywood News. He has written two other novels: "Blackberries Got No Thorns" and "The Voodoo Vortex." He resides with his wife in Memphis, Tennessee. Reach him through email at Rumor3@aol.com or his website www.rusmor.com

www.ingramcontent.com/pod-product-compliance
Lightning Source LLC
Chambersburg PA
CBHW032001170526
45157CB00002B/499

* 9 7 8 1 4 0 3 3 5 4 9 8 3 *